HYBRID FINITE ELEMENT METHOD FOR STRESS ANALYSIS OF LAMINATED COMPOSITES

HYBRID FINITE ELEMENT METHOD FOR STRESS ANALYSIS OF LAMINATED COMPOSITES

by

Suong Van Hoa
Concordia University

and

Wei Feng
Concordia University

SPRINGER SCIENCE+BUSINESS MEDIA, LLC

Library of Congress Cataloging-in-Publication Data

A C.I.P. Catalogue record for this book is available
from the Library of Congress.

ISBN 978-1-4613-7626-2 ISBN 978-1-4615-5733-3 (eBook)
DOI 10.1007/978-1-4615-5733-3

Printed on acid-free paper.

CONTENTS

PREFACE

This book has one single purpose: to present the development of the partial hybrid finite element method for the stress analysis of laminated composite structures. The reason for this presentation is because the authors believe that partial hybrid finite element method is more efficient that the displacement based finite element method for the stress analysis of laminated composites. In fact, the examples in chapter 5 of this book show that the partial hybrid finite element method is about 5 times more efficient than the displacement based finite element method. Since there is a great need for accurate and efficient calculation of interlaminar stresses for the design using composites, the partial hybrid finite method does provide one possible solution.

Hybrid finite method has been in existence since 1964 and a significant amount of work has been done on the topic. However, the authors are not aware of any systematic piece of literature that gives a detailed presentation of the method. Chapters 1 and 2 present a summary of the displacement finite element method and the evolution of the hybrid finite element method. Hopefully, these two chapters can provide the readers with an appreciation for the difference between the displacement finite element method and the hybrid finite element. It also should prepare the readers for the introduction of partial hybrid finite element method presented in chapter 3. In this chapter, a new composite variational principle is presented along with the formulation of the partial hybrid finite element. One of the most trouble aspect of the hybrid finite element method is the determination of the stress modes. Section 3.4 provides three different methods for systematic determination of the stress modes. In chapter 4, many partial hybrid finite elements are introduced. These include the single layer finite elements and the multilayer finite elements. Among the two groups of these elements, there are solid element for the local region, the laminated element for the global region and the transition element for the transition region. These elements are useful for the application of the global/local approach used in chapter 5. Many examples to illustrate the efficiency of the partial hybrid finite element method are presented in chapter 5. These include problems with free edge effects such as laminate with straight edges subject to uniaxial extension and laminates with holes and subjected to in plane tension. These results show that the partial hybrid elements can provide accurate interlaminar stresses with strong efficiency.

This work is the culmination of three generations of Ph.D students in the

Concordia Center for Composites. The authors would like to thank Dr. Q. Huang and Dr. J. Han for their previous works which form the important contributions to the book. The second author is grateful to Professors Wei-zang Chien and Qian Huang of Shanghai University for introducing him to generalized variational principles and composite materials.

A work of this magnitude requires strong support from family members. The first author would like to thank his mother, Hoa, Thi Tho for her dedication and sacrifice during his early years; his wife, Dong thi Do for her love, dedication and support; and his four children Vincent, Glenn, Sabrina and Victoria for their inspiration. The second author wishes to thank his mother, Feng-zhen Sun, his father, Sui-gen Feng, his mother in-law, Jin-yin Zhu and his father in-law, Yun-long Zhang for their solid support. He also thanks his wife, Pei-jun Zhang and his daughter, Fan Feng for their moral support and understanding.

HYBRID FINITE ELEMENT METHOD FOR STRESS ANALYSIS OF LAMINATED COMPOSITES

Chapter 1

INTRODUCTION

1.1 INTRODUCTION

When one talks about doing analysis of composites, one tends to mean the analysis of long continuous fibre laminated composites. This is because, while other forms of composites (such as short fiber composites) do exist and are important, they can be adequately assumed to be isotropic for analysis purposes, and the analysis (stress analysis, vibration, buckling) of structures made of these materials can be proceeded in a manner similar to the case of isotropic materials. The analysis of long continuous fiber reinforced composites, on the other hand, presents additional complication. For example, there is the anisotropy due to fiber orientation and the state of stress depends on the stacking sequence of the laminated composites, the fibre orientation of each lamina as well as the material properties of the fibre and of the matrix. Therefore, finite element method is widely used in the analysis of structures made of long continuous fiber reinforced composites. This is due to the power of the technique to be able to model the laminated composite structures not only in the planer dimensions, but also in the thickness direction. It is also due to the availability of many commercial finite element codes such as ALGOR, ANSYS, MSC/NASTRAN, PATRAN3, and so on. The finite element methods as discussed in this book are restricted to the analysis of structures made of long continuous fibre composite materials.

Analysis of structures made of composites (or for that matter of structures made of any material) can be classified into three categories. These are the stress analysis, vibration analysis and instability (buckling) analysis. In vibration and buckling analyses, eigenvalues are solved, eigenvectors are extracted and engineering quantities such as vibration frequencies and/or critical buckling loads are derived. For these analyses, the overall stiffness of the structure is important and the analyst may afford to model the composite structure with a small number of fairly large elements to obtain accurate results. The purpose of stress analysis, on the other hand, is to provide stress or strain values so that failure or residual life can be predicted. Since failure of a structure is determined by the failure of its weakest link, it is important to determine accurately the stresses at the critical locations in the composite structures. These analyses may require a large number of elements and a fine element mesh. Different types of the analysis of composite structures therefore require different types of finite elements and element meshes.

There are a few elements which can be chosen in the commercial finite element programs for analysis of structures made of composites. Finite elements in different programs are based on different theories. The power of the programs depends essentially on the basic theory formulating the element. In general, composite structures are modelled using one of the following two classes of theories[1.1-1.2]:

1. Equivalent single-layer 2-D theories[1.3-1.8], in which deformable models are based on global through-the-thickness displacement, strain and stress approximations;

2. 3-D continuum theories[1.9-1.10], in which each of the individual layers of a composite structure is treated as a three-dimensional continuum.

Corresponding to two classes of theories above, the finite element models can be classified into three classes: laminated elements based on the equivalent single-layer 2-D theories; 3-D solid elements and multilayer elements based on the 3-D continuum theories.

Laminated Elements

The close to exact modelling of laminated composites requires the discretization not only along the surface of the composite structure (usually plate or shell) but also across the thickness of the structure (across the different layers). This discretization can result in a large number of elements which translates into large requirement in computer space and time. In addition, the aspect ratios in the elements can become excessive. Large aspect ratios may create problems in shear locking. Due to these problems, the detailed and thorough modelling of the composite is usually avoided if possible. Fortunately, this can be done without much sacrifice in accuracy in problems where the overall stiffness of the structure is important and the detailed and accurate stresses (particularly interlaminar stresses) at different points in the structure are not of concern. Problems such as the determination of vibration frequencies and mode shapes, the determination of critical buckling loads and buckling modes fall into this category. For the analysis of these problems, the stiffness of the laminate is obtained by integration of the stiffness of individual layers across the thickness. The thickness of each finite element is the same as the thickness of the laminate. By smearing the thickness of the individual layers, significant reduction in computer requirements can be achieved. This type of elements is called as the "laminated element" [1.5]. In the laminated elements, therefore, the variation in fibre orientations and material properties across the thickness is integrated to obtain a single property across the thickness. As indicated, this element can be used for problems such as vibration or buckling but does not provide useful results if interlaminar stresses are required.

3-D Solid Elements and Multilayer Elements

For problems where the analyst is interested in the prediction of failure of the composite structure, it is important to know the stresses (particularly interlaminar stresses) accurately. As such, the composite structure needs to be modelled in great details, particularly at locations where there is suspicion of large stress gradients. Thus, 3-D solid elements and multilayer elements are usually used for these problems. In 3-D solid elements [1.11], no specific kinematic assumptions are introduced regarding the behaviour of a laminated composite. Across the thickness of the laminated composite, each layer is modelled using one or more elements. It takes the behaviour of the individual laminae into consideration. In the multilayer elements [1.12], the individual laminae are modeled using one or more 3-D sub-elements. These sub-elements are then assembled through the thickness according to the continuity conditions on displacements and stresses. In order to minimise the problem of large aspect ratios, the planar dimensions of the elements should be kept to be not too large compared to the thickness of the element. Because composite laminae are very thin and a typical laminated composite may contain many laminae, this usually results in an excessively large number of elements which means large requirements in computer space and time.

Many problems such as delamination and splitting, which are the common modes of failure in composites, require determination of accurate stresses, particularly interlaminar stresses. However, the limitation of computer capacity limits the ability of composite analysts to obtain a good handle in the prediction of failure. The solution to this problem can be done from two directions: the development of more powerful computers (within the affordable costs) which depends on the capacity of computer scientists and engineers, and/or the development of more efficient finite elements for the stress analysis of laminated composites.

The task of developing good finite elements never seems to be finished for the stress analysis of laminated composites. Today, the majority of finite element analysis for composite structures is using single-field finite elements that are formulated based on displacement. This is due to the simple approach to the element formulation provided by displacement finite element model. The displacement finite elements work well with the stress analysis of homogenous materials. However, they can not satisfy well the requirements in analysis of laminated composites. As an alternate, the conventional hybrid stress elements [1.13-1.17] have been used to analyze them. In order to show the advantages and disadvantages of displacement elements, this chapter will briefly present the finite element method based on displacement formulation, and then, it will give the motivation for developing hybrid finite elements. The basic knowledge of the theories of laminated composite is referred to the literature [1.18-1.20].

1.2 DISPLACEMENT FINITE ELEMENT METHOD

Finite element method is an approximate technique. This means that the continuous structure is discretized into a number of continuous elements connected together at a number of nodes. As the number of elements increases, the approximation of the structure becomes more and more accurate. The solution obtained from finite element method therefore is an approximate solution. As the number of elements increases, the general tendency is that the finite element solution also approaches the exact solution.

The finite element procedure for stress analysis of structures is a systematic numerical approximation which can be implemented on a computer. Its generality fits the analysis requirements of today's complex engineering systems and designs where closed-form solutions of governing equations are usually not available. In general, finite element analysis of structures is performed by following six steps: discretizing the structure, deriving element equations, assembling elements, imposing essential boundary conditions, solving primary unknowns, and calculating secondary quantities.

Discretizing the Structure

First of all, in order to analyze a structure using finite element method, the structure must be discretized into a suitable number of "small" bodies, called "finite elements", which are quasi-disjoint non-overlapping elements. These elements are connected by using a set of key points, called "nodes" (see figures 1 and 2).

How to discretize a structure or how many elements to be used depends on the problem to be analyzed. In general, the region where large gradients of displacements and/or stresses to be expected is discretized into many elements (fine mesh); otherwise, the region is modeled using few elements (coarse mesh). For example, if a structure contains a crack or a open hole, the local region near the crack or the hole is usually divided into many "very small" elements in order to predict accurately the response of the structure. In addition, what type of element to be used depends on the characteristics of the continuum. For laminated composite, plate/shell elements are commonly used to predict vibration frequencies and critical buckling loads of structures, and 3-D solid elements and multilayer elements are usually used to predict stresses in the structures.

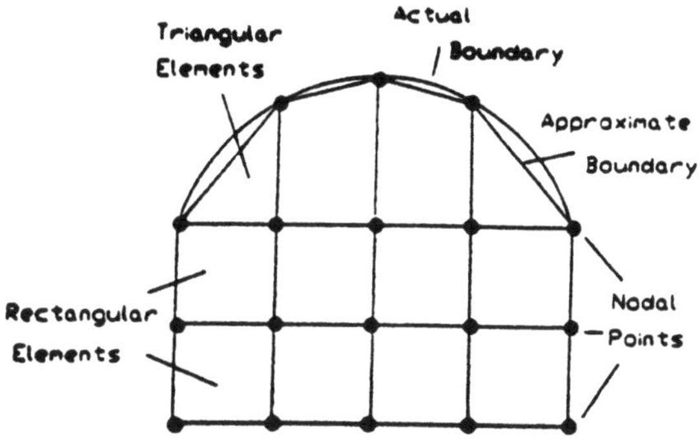

Figure 1 Plane stress problem

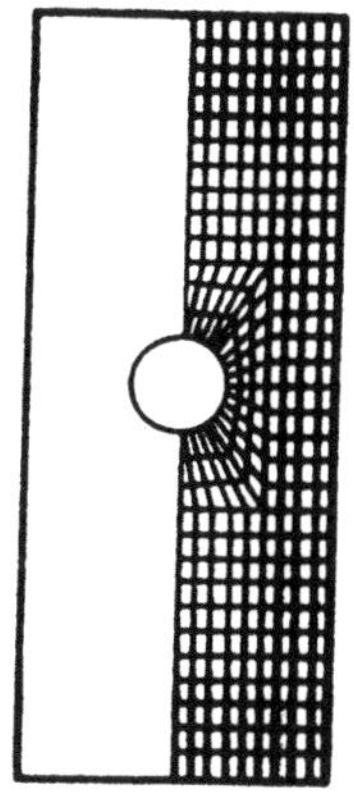

Figure 2 A plate with a hole

Deriving Element Equations

Once the mesh of structures is generated and the type of elements is determined, element equations (mass matrix, stiffness matrix, nodal loading vector, etc.) can be derived. Four methods are available to derive element matrices and equations: the direct method, the variational method, the weighted residual method, and the energy method. In this book, the variational method for deriving the element equation is presented.

The variational principle is stated as $\delta\Pi_p=0$ [1.21] with

$$\Pi_p = U - \int_{V_e} \mathbf{F}^T \mathbf{u}\, dV - \int_{S_e} \mathbf{T}^T \mathbf{u}\; dS \tag{1-1}$$

where **u** is the displacement field. **F** and **T** are respectively the prescribed body forces in the domain V_e and boundary traction along the boundary S_σ. In the equation, the strain energy U is

$$U = \int_{V_e} \frac{1}{2} \boldsymbol{\varepsilon}^T [C]\; \boldsymbol{\varepsilon}\, dV \tag{1-2}$$

where the strains ε must satisfy the compatibility equation, i.e the strain displacement relation.

In the finite element approach, when the domain V to be analyzed is decomposed into a finite number of non-overlapping subdomains V_e which are called the elements, these elements will be interconnected at a finite number of points called the nodal points. In an element, therefore, the displacement **u** is described by the nodal displacement δ_e

$$\mathbf{u} = [N]\,\boldsymbol{\delta}_e \tag{1-3}$$

where [N] is the displacement shape function. Using the strain displacement relation, the strains can then be computed in terms of the nodal displacements as,

$$\boldsymbol{\varepsilon} = [B]\,\boldsymbol{\delta}_e \quad \text{and} \quad \varepsilon_{ij} = \frac{1}{2}(u_{i,j} + u_{j,i}) \tag{1-4}$$

where [B] is the geometry matrix. Substituting equations (1-2), (1-3) and (1-4) into equation (1-1), the variational functional can be rewritten as

$$\begin{aligned} \Pi_p = {} & \frac{1}{2}\boldsymbol{\delta}_e^T\left(\int_{V_e}[B]^T[C]\,[B]\,dV\right)\boldsymbol{\delta}_e \\ & -\boldsymbol{\delta}_e^T\left(\int_{V_e}[N]^T\mathbf{F}\,dV + \int_{S_\sigma}[N]^T\mathbf{T}\,dS\right) \end{aligned} \tag{1-5}$$

Denote

$$\begin{aligned} [K]_e &= \int_{V_e}[B]^T[C]\,[B]\,dV \\ \mathbf{f}_e &= \int_{V_e}[N]^T\mathbf{F}\,dV + \int_{S_\sigma}[N]^T\mathbf{T}\,dS \end{aligned} \tag{1-6}$$

where [C] is the material constant matrix, $[K]_e$ is usually referred to as the element stiffness matrix and $\mathbf{f}_e$ is the equivalent nodal force vector. Thus, the governing equation of the element is obtained by using $\delta\Pi_p=0$,

$$[K]_e\,\boldsymbol{\delta}_e = \mathbf{f}_e \tag{1-7}$$

By means of this formulation, equations of all displacement elements in the finite element model can be derived.

Assembling Elements

Once the element equations (1-7) are established, the global equations that define approximately the behavior of the structure can be assembled by means of the continuity conditions between adjacent elements. For example, the displacements of two adjacent points in the finite element model must have identical values. Algebraically, the ranges of indices on element matrices $[K]_e$ and $\mathbf{f}_e$ can be expanded to the total number of degrees of freedom of the structure. It is equivalent to adding rows and columns of zeros in $[K]_e$ and $\mathbf{f}_e$ for all degrees of freedom which are not contained in the element. Thus, the assembly of the global stiffness matrix and load vector is accomplished by the summation of element contributions as

$$[K] = \sum_{i=1}^{N} [K]_i^e \qquad \mathbf{f} = \sum_{i=1}^{N} \mathbf{f}_i^e \tag{1-8}$$

where N stands for the number of elements. Finally, the global equations are obtained and can be expressed in matrix notation as

$$[K]\,\boldsymbol{\delta} = \mathbf{f} \tag{1-9}$$

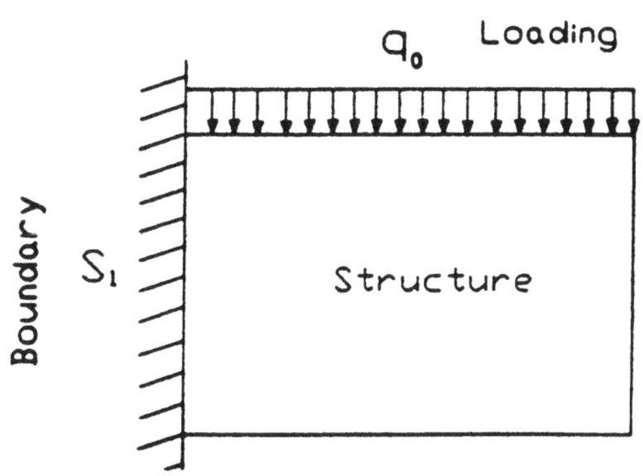

Figure 3 Prescrbing displacements on boundary S_1

Imposing Boundary Conditions

Boundary conditions are the physical constraints or supports that must exist so that the structure can stand in space uniquely. These conditions are commonly specified in terms of known values of the nodal displacements on a part of surface or boundary S_1 (see figure 3). The global stiffness matrix [K] and load vector **f** are modified by applying boundary conditions to produce the final global matrix [K] and vector **f**. A simple way is to replace the equations for nodes on the boundary by the prescribed nodal value and modify the other equations accordingly.

Solving Primary Unknowns

The global equations (1-9) are a set of linear algebraic equations, which can be expressed in the form as

$$\begin{aligned} K_{11}u_1+K_{12}u_2+\ldots+K_{1n}u_n&=f_1 \\ K_{21}u_1+K_{22}u_2+\ldots+K_{2n}u_n&=f_2 \\ &\vdots \\ K_{n1}u_1+K_{n2}u_2+\ldots+K_{nn}u_n&=f_n \end{aligned} \tag{1-10}$$

where, the unknown nodal displacement u_i (i=1,2,..,n) are called primary unknowns because they appear as the first quantities sought in the basic equations (1-10). There are many methods to solve the set of equations, such as Gaussian elimination and iterative methods. The choice of appropriate solution algorithms will affect the overall efficiency of the computation.

Calculating Secondary Quantities

Secondary quantities must be computed from primary quantities. In solid mechanics, such quantities can be strains and stresses. They are calculated from the displacement by means of numerical differentiation. With the strains

$$\boldsymbol{\varepsilon}=[B]\,\boldsymbol{\delta}_e \tag{1-11}$$

the stresses

$$\boldsymbol{\sigma} = [C]\,\boldsymbol{\varepsilon} \qquad (1\text{-}12)$$

can be computed when nodal displacement δ_e are known. Matrix [B] is a function of the coordinates and must be evaluated at the point where stresses are desired. In the displacement finite element models, the displacements are guaranteed to be continuous across the boundary of adjacent elements, but the strain and hence stress continuity across the boundary is not guaranteed. They may take different values on the two sides of the interface between elements. In order to satisfy the continuity condition of stresses at the interface, hybrid elements are introduced. This will be discussed later.

1.3 ASSUMED DISPLACEMENT FIELD

For finite element computation, the approximation space, a space of finite dimension, is generated from element basis functions. For instance, an approximate displacement solution can be expressed in the form,

$$u_i = \sum_{j=1}^{m} a_{ij} g_j(x,y,z) \qquad (i=1,2,3) \qquad (1\text{-}13)$$

where $g_j(x,y,z)$ are element basis functions. Because these functions limit the infinite degrees of freedom of the system, the properties of these functions determine the character of the finite element approximation space. In order to ensure convergence to the correct result, three simple requirements have to be satisfied [1.22].

<u>Requirement 1</u>:

> The displacement function chosen should be such that, when the nodal displacements are caused by a rigid body displacement, the element is able to reproduce the rigid body motion.

<u>Requirement 2</u>:

> The displacement function chosen should be such that, if nodal displacements are compatible with a constant strain condition, the element is able to reproduce the constant strain deformation.

<u>Requirement 3</u>:

> The displacement function chosen should be such that the strains at the interface between adjacent elements are finite.

A number of mathematical functions such as trigonometric series and exponential functions can be used as element basis functions. However, orthogonal polynomials are more appropriate as basis functions because of the ease and simplification they provide in the finite element formulation. The choice of polynomial depends on the type of element used.

Compatible Elements

In the regular finite element method in solid mechanics, a compatible (conforming) displacement field is often used as the dependent variable. In this case, the number of terms in the polynomial must be equal to the total number of degrees of freedom associated with the element, otherwise the polynomial may not be unique. Thus, for a bar element with two nodes, one degree of freedom at each node (see figure 4a), the displacement can be assumed in the form of a two-term polynomial

$$u_1 = a_1 + a_2\, x \qquad (1 - 14)$$

For a triangular element with three nodes in a two dimensional problem, two degrees of freedom at each node (see figure 4b), the displacement can be assumed in the form of two three-term polynomials

$$\begin{aligned} u_1 &= a_1 + a_2\, x + a_3\, y \\ v_1 &= b_1 + b_1\, x + b_3\, y \end{aligned} \qquad (1 - 15)$$

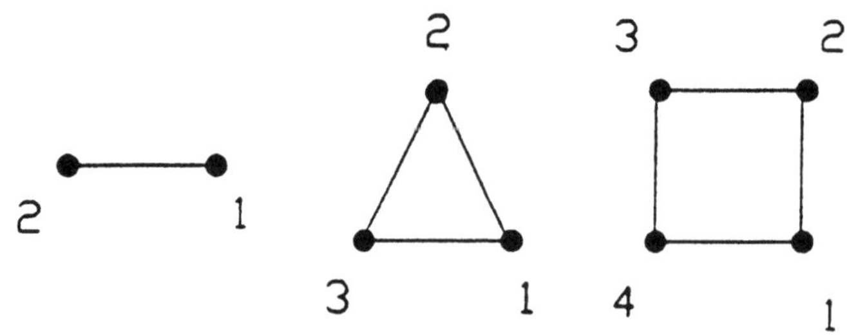

(a) (b) (c)

Figure 4 1-D and 2-D linear elements

Furthermore, for a rectangular element with four nodes in a two dimensional problem, two degrees of freedom at each node (see figure 4c), the displacement can be assumed in the form of two four-term polynomials

$$u_1 = a_1 + a_2\, x + a_3\, y + a_4\, x\, y$$
$$v_1 = b_1 + b_2\, x + b_3\, y + b_4\, x\, y \qquad (1 - 16)$$

where the coefficients a_i and b_i are called generalized variables. For the polynomial series, the three requirements for convergence above can be met by satisfying the continuity and completeness conditions.

Continuity means that these functions and their derivatives, where required, must be continuous within the element domain and across the interface between adjacent elements. The linear function (eq. 1-14) is indeed continuous within the elements. For compatibility, continuity condition across the interface between adjacent elements, it is necessary that the coordinates and displacements of the elements at the interface be the same. Because the coordinates and displacements of an element on the interface are determined only by nodes and nodal degrees of freedom on that interface, compatibility is satisfied if the adjacent elements have the same nodes on the interface and the coordinates and displacements on the interface are defined in each element by the same polynomial functions (see figure 5 a and b).

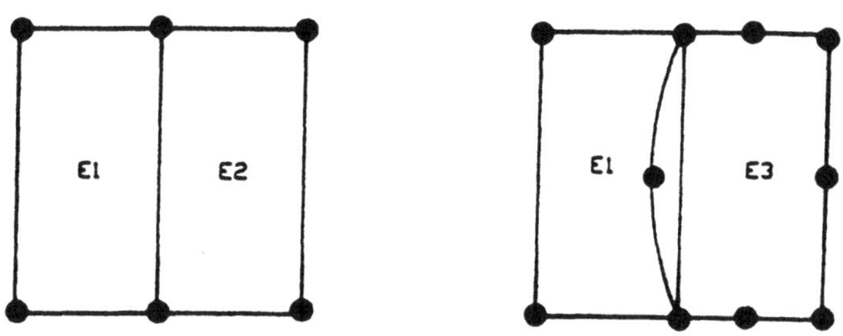

(a) Continuous displacements (b) Discontinuous displacement

Figure 5 Compatibility condition between elements

Completeness means that the polynomial functions must contain the constant and linear terms so that the element nodes can be given rigid body displacements without producing strain within the element and can always recover state of constant strain. For instance, the linear approximation (eq. 1-14) contains the constant term a_1 which allows for the rigid body displacement mode. Also, in one dimensional problem, the linear approximation (eq. 1-14) contains linear term a_2x which guarantees that the element is able to recover the state of constant strain. This condition implies that, when the element becomes smaller and smaller, the strain in the element approaches a constant value.

The necessary terms for a complete polynomial are presented by Pascal's triangle which is shown below.

$$
\begin{array}{ccccccccc}
 & & & & 1 & & & & \\
 & & & x & & y & & & \\
 & & x^2 & & xy & & y^2 & & \\
 & x^3 & & x^2y & & xy^2 & & y^3 & \\
x^4 & & x^3y & & x^2y^2 & & xy^3 & & y^4
\end{array}
$$

Thus, a complete quadratic polynomial is of the form

$$a_1 + a_2x + a_3y + a_4x^2 + a_5xy + a_6y^2 \tag{1-17}$$

and requires an element with six degrees of freedom to uniquely define a_i. Moreover, a complete cubic polynomial is of the form

$$\begin{gathered} a_1 + a_2x + a_3y + a_4x^2 + a_5xy + a_6y^2 \\ + a_7x^3 + a_8x^2y + a_9xy^2 + a_{10}y^3 \end{gathered} \tag{1-18}$$

and requires an element with ten degrees of freedom to uniquely define a_i. Furthermore, the completeness condition also requires to use the least-order terms in displacement basis function.

Non-compatible Elements and the Patch Test

Compatible elements are always desired in finite element analysis. But in some cases, there is considerable difficulty in finding displacement basis functions for an element which guarantees that displacements are continuous on the interface between adjacent elements. The discontinuity of displacements will cause infinite strains at the interface. However, if a non-compatible (nonconforming) element can pass a test, called patch test, the finite element solution will still tend to the correct answer when the size of elements tends to be small and the element mesh of a structure becomes very fine. The detail discussion of non-compatible elements is beyond this book and is referred to the literature [1.22].

The patch test was first introduced by Irons [1.23]. To do a "patch test", one assembles a small number of elements into a "patch". The meshes suitable for patch test calculations in 2-D patch are shown in figure 6. Then, the nodal forces corresponding to a series of constant stress states are applied to boundary nodes of the patch. If, computed stresses in the element always agree with expectation, the patch test is passed. If an element passes the patch test, the solution of the finite element model using this type of element will converge to a correct result. Today, the patch test serves as a necessary and sufficient condition for correct convergence of a finite element formulation.

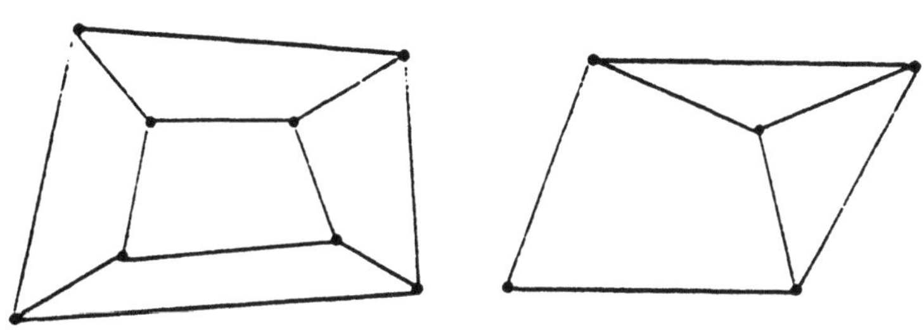

Figure 6 The element meshes for patch test

Isoparametric Elements and Numerical Integration

Isoparametric elements first appeared in the literature in 1966 [1.24]. In the displacement finite element, there are two entities that need to be approximated. The first is physical (the displacement field) and the second is geometrical (the shape of the element). Therefore, it must be decided whether to approximate physics and geometry equally or to give preference to one or the other in the element. For isoparametric displacement elements, both element geometric shape and displacement interpolation polynomial are required to be mapped from the global coordinate system to the parametric coordinate system (see figure 7a and b). The term "isoparametric", meaning "same parameters", follows from use of the same interpolation polynomial to define both the geometry and the displacement field of an element. Thus, the coordinates of a point within an element and the displacement of the element can be expressed in the same form,

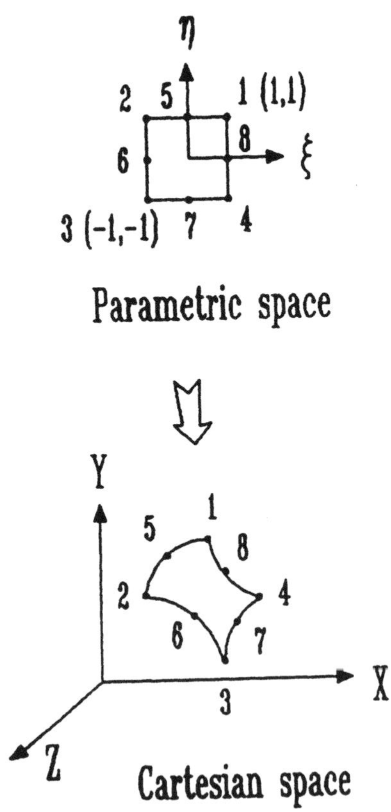

Figure 7a Two-dimensional isoparametric element

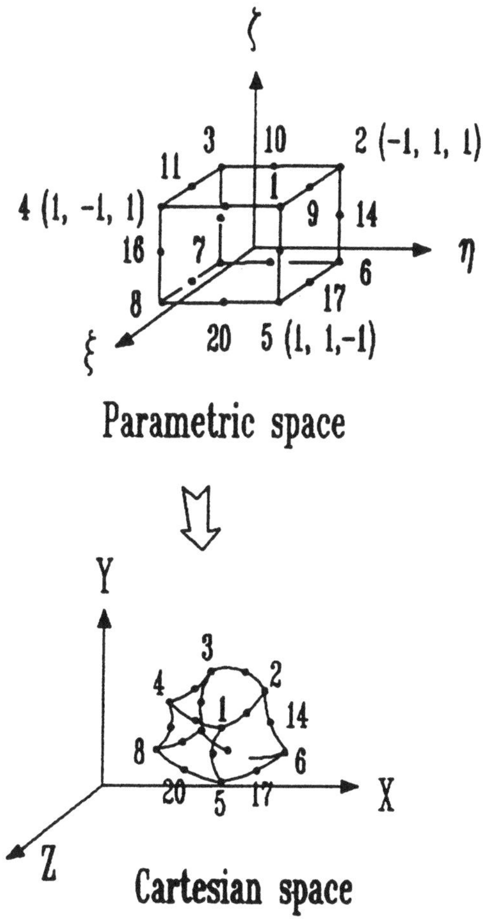

Figure 7b Three-dimensional isoparametric element

$$x=\sum_{i=1}^{m} N_i(\xi,\eta,\zeta)\,x_i \quad y=\sum_{i=1}^{m} N_i(\xi,\eta,\zeta)\,y_i \quad z=\sum_{i=1}^{m} N_i(\xi,\eta,\zeta)\,z_i$$
$$u=\sum_{i=1}^{m} N_i(\xi,\eta,\zeta)\,u_i \quad v=\sum_{i=1}^{m} N_i(\xi,\eta,\zeta)\,v_i \quad w=\sum_{i=1}^{m} N_i(\xi,\eta,\zeta)\,w_i \tag{1-19}$$

Equation (1-19) allows elements with curved sides because the element sides are fitted between the nodes. In standard-type element formulation, this is not possible because the element sides are always straight regardless of any mid-nodes. The shape functions N_i for various elements can be found in the literature [1.25]. For examples, in two dimensional problem, the shape functions of a linear element (see figure 8) are

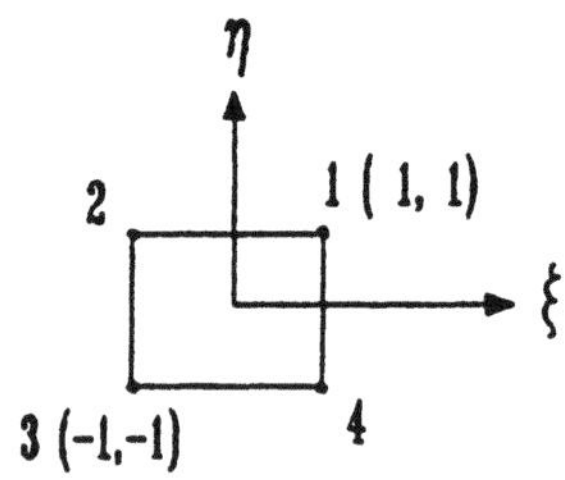

(a) Linear element

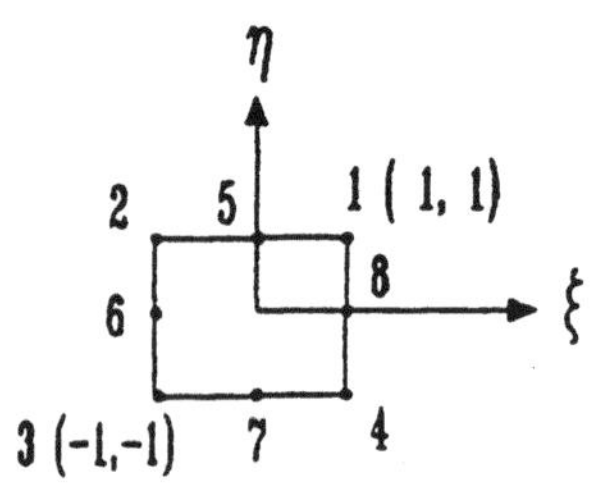

(b) Quadratic element

Figure 8 Two-dimensional elements

$$N_i=\frac{1}{4}(1+\xi_0)(1+\eta_0) \tag{1-20}$$

in which,

$$\xi_0=\xi_i\xi \quad \eta_0=\eta_i\eta \quad (i=1,2,3,4) \tag{1-21}$$

where ξ_i, and η_i are the local co-ordinates of node i in the element parametric space. The shape functions of a quadratic element (see figure 8) in two dimensional problem are

$$\begin{aligned} N_i=&\frac{1}{4}(1+\xi_0)(1+\eta_0)(\xi_0+\eta_0-1)\xi_i^2\eta_i^2 \\ &+\frac{1}{2}(1-\xi^2)(1+\eta_0)(1-\xi_i^2)\eta_i^2 \\ &+\frac{1}{2}(1-\eta^2)(1+\xi_0)(1-\eta_i^2)\xi_i^2 \end{aligned} \tag{1-22}$$

in which,

$$\xi_0=\xi_i\xi \quad \eta_0=\eta_i\eta \quad (i=1,2,\ldots,8) \tag{1-23}$$

In three dimensional problem, the shape functions of a linear element (see figure 9) are

$$N_i=\frac{1}{8}(1+\xi_0)(1+\eta_0)(1+\zeta_0) \tag{1-24}$$

in which,

$$\xi_0=\xi_i\xi \quad \eta_0=\eta_i\eta \quad \zeta_0=\zeta_i\zeta \quad (i=1,2,\ldots,8) \tag{1-25}$$

where ξ_i, η_i and ζ_i are the local co-ordinates of node i in the element parametric space. The shape functions of a quadratic element (see figure 9) are

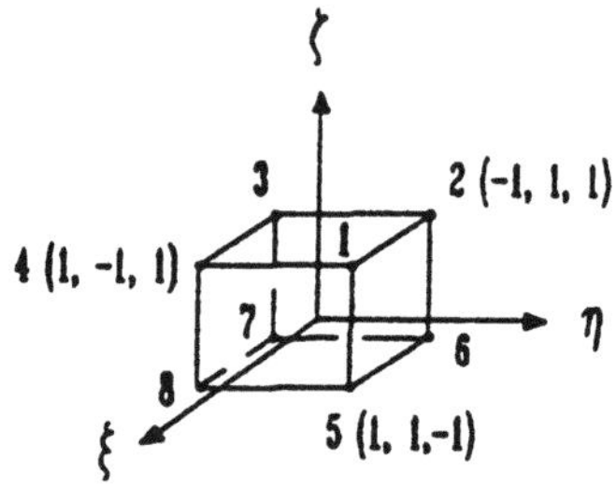

(a) Linear element

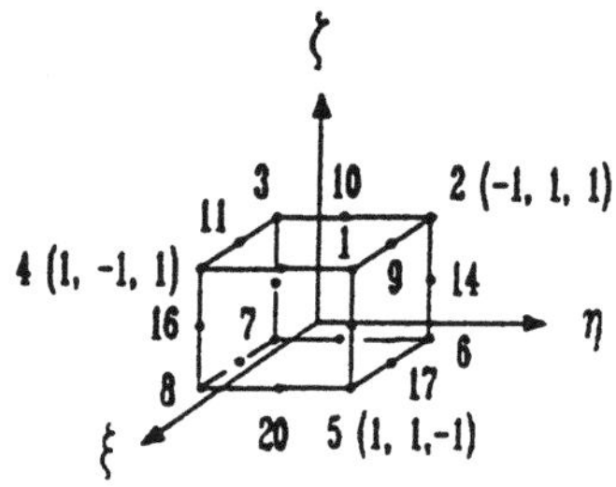

(b) Quadratic element

Figure 9 Three-dimensional elements

$$
\begin{aligned}
N_i = & \frac{1}{8}(1+\xi_0)(1+\eta_0)(1+\zeta_0)(\xi_0+\eta_0+\zeta_0-2)\,\xi_i^2\eta_i^2\zeta_i^2 \\
& +\frac{1}{4}(1-\xi^2)(1+\eta_0)(1+\zeta_0)(1-\xi_i^2)\,\eta_i^2\zeta_i^2 \\
& +\frac{1}{4}(1-\eta^2)(1+\zeta_0)(1+\xi_0)(1-\eta_i^2)\,\zeta_i^2\xi_i^2 \\
& +\frac{1}{4}(1-\zeta^2)(1+\xi_0)(1+\eta_0)(1-\zeta_i^2)\,\xi_i^2\eta_i^2
\end{aligned}
\tag{1-26}
$$

in which,

$$\xi_0=\xi_i\xi \quad \eta_0=\eta_i\eta \quad \zeta_0=\zeta_i\zeta \qquad (i=1,2,\ldots,20) \tag{1-27}$$

In order to perform the evaluation of isoparametric element matrices, a coordinate transformation of derivatives is required because the displacements are given in terms of parametric coordinates ξ, η, and ζ. Therefore, the Jacobian matrices must be calculated. For two dimensional problem, the jacobian matrix is

$$[J]=\begin{bmatrix}\frac{\partial x}{\partial \xi} & \frac{\partial y}{\partial \xi}\\ \frac{\partial x}{\partial \eta} & \frac{\partial y}{\partial \eta}\end{bmatrix} \tag{1-28}$$

The derivatives of displacements are written in the form

$$\begin{Bmatrix}\frac{\partial u_i}{\partial x}\\ \frac{\partial u_i}{\partial y}\end{Bmatrix}=[J]^{-1}\begin{Bmatrix}\frac{\partial u_i}{\partial \xi}\\ \frac{\partial u_i}{\partial \eta}\end{Bmatrix} \qquad (i=1,2) \tag{1-29}$$

The element stiffness matrix is

$$[K]_e=\int_{V_e}[B]^T[C][B]\,t\,dx\,dy=\int_{-1}^{1}\int_{-1}^{1}[B]^T[C][B]\det[J]\,t\,d\xi\,d\eta \tag{1-30}$$

where t is the thickness of the element. For three dimensional problem, the Jacobian matrix is

$$[J]=\begin{bmatrix}\frac{\partial x}{\partial \xi} & \frac{\partial y}{\partial \xi} & \frac{\partial z}{\partial \xi}\\ \frac{\partial x}{\partial \eta} & \frac{\partial y}{\partial \eta} & \frac{\partial z}{\partial \eta}\\ \frac{\partial x}{\partial \zeta} & \frac{\partial y}{\partial \zeta} & \frac{\partial z}{\partial \zeta}\end{bmatrix} \tag{1-31}$$

and the transformation of derivatives and stiffness matrix is similar to the two dimensional case. The integral in equation (1-30) is evaluated numerically using Gauss quadrature. Sampling points are required to evaluate integrals numerically by using Gauss quadrature. The number of sampling points used to evaluate element integrals is given in Table 1.

Table 1 Two-dimensional Gauss quadrature order

Number of nodes	Element shape	Reliable Gauss quadrature order	Reduced order
4		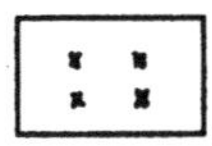 2×2	Same
8		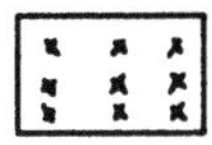 3×3	2×2
12		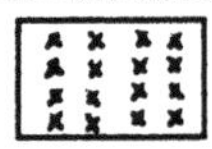 4×4	3×3

1.4 DISPLACEMENT ELEMENT FORMULATION

As mentioned above, there are three classes of displacement finite elements for the analysis of composite structures: laminated elements, 3-D solid elements, and multilayer elements. The element matrices and equations of these elements can be obtained by using equations (1-6) and (1-7).

Laminated Plate/Shell Element

One way to derive element formulation for the behaviour of plate/shell is to apply specific kinematic constraints to the full three-dimensional elasticity equations. This 'degeneration' of the three-dimensional elasticity equations is the basis for many plate/shell formulations. Based on the kinematic assumptions, the number of displacement parameters through the thickness can be significantly reduced compared to 3-D modelling.

A 4-node degenerated plate element (see figure 10a) is presented as follows. Firstly, the global co-ordinates (x,y,z) of any point within the element can be expressed in the form specified by the 'vector' connecting the upper and lower points (see figure 10b) and the mid-surface co-ordinates as

$$\begin{Bmatrix} x \\ y \\ z \end{Bmatrix} = \sum_{i=1}^{4} N_i(\xi,\eta) \begin{Bmatrix} x_i \\ y_i \\ z_i \end{Bmatrix} + \sum_{i=1}^{4} N_i(\xi,\eta)\, h_i\, \frac{\zeta}{2} \mathbf{V}_{3i} \tag{1-32}$$

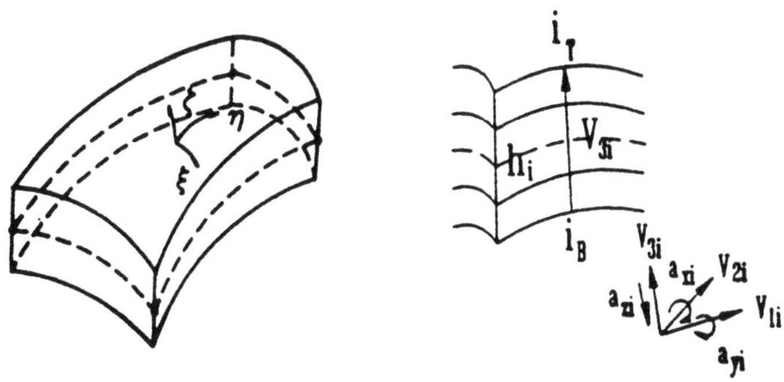

(a) (b)

Figure 10 A degenerated plate/shell element

where

$$\mathbf{V}_{3i} = \begin{Bmatrix} l_{3i} \\ m_{3i} \\ n_{3i} \end{Bmatrix} = \frac{1}{h_i} \left(\begin{Bmatrix} x_i \\ y_i \\ z_i \end{Bmatrix}_T - \begin{Bmatrix} x_i \\ y_i \\ z_i \end{Bmatrix}_B \right) \tag{1-33}$$

and

$$h_i = \sqrt{(x_{iT} - x_{iB})^2 + (y_{iT} - y_{iB})^2 + (z_{iT} - z_{iB})^2} \tag{1-34}$$

The $N_i(\xi,\eta)$ are shape functions, ξ and η are the normalized curvilinear co-ordinates in the middle plane of the plate/shell, ζ is a linear co-ordinate in the thickness direction and only approximately normal to the mid-surface initially, and (x_i, y_i, z_i) are the global co-ordinates at node i. The shape functions are

$$N_i=\frac{1}{4}(1+\xi_0)(1+\eta_0) \tag{1-35}$$

in which,

$$\xi_0=\xi_i\xi \qquad \eta_0=\eta_i\eta \qquad (i=1,2,3,4) \tag{1-36}$$

In the element, the displacement field is assumed as a continuous field through the entire laminate thickness. It is also assumed that a line that is straight and normal to the middle surface before deformation is still straight, but not necessarily 'normal' to the middle surface after deformation. The displacement throughout the element will be uniquely defined by three Cartesian components (u_i, v_i and w_i) of the displacement of the mid-surface node i, two rotations (a_{xi} and a_{yi}) of the nodal vector $\mathbf{V}_{3i}$ about orthogonal directions normal to it, and one transverse normal deformation (a_{zi}) in the thickness direction.

$$\mathbf{u}_i=\begin{Bmatrix} u_i \\ v_i \\ w_i \end{Bmatrix}+\frac{h_i}{2}\zeta\,[\mathbf{V}_{1i} \;\; -\mathbf{V}_{2i} \;\; \mathbf{0}]\begin{Bmatrix} a_{xi} \\ a_{yi} \\ 0 \end{Bmatrix}+\frac{h_i}{2}\zeta\,[\mathbf{0} \;\; \mathbf{0} \;\; \mathbf{V}_{3i}]\begin{Bmatrix} 0 \\ 0 \\ a_{zi} \end{Bmatrix} \tag{1-37}$$

in which, $\mathbf{V}_{1i}$, $\mathbf{V}_{2i}$ and $\mathbf{V}_{3i}$ are the unit vectors of the local co-ordinate (ξ, η, ζ) at node i. They can be calculated as follows:

$$\mathbf{V}_{1i}=\begin{Bmatrix} l_{1i} \\ m_{1i} \\ n_{1i} \end{Bmatrix}=\frac{\mathbf{i}\times\mathbf{V}_{3i}}{|\mathbf{i}\times\mathbf{V}_{3i}|} \qquad \mathbf{V}_{2i}=\begin{Bmatrix} l_{2i} \\ m_{2i} \\ n_{2i} \end{Bmatrix}=\mathbf{V}_{3i}\times\mathbf{V}_{1i} \tag{1-38}$$

If $\mathbf{i}\times\mathbf{V}_{3i}=0$, $\mathbf{i}$ can be replaced by $\mathbf{j}$. Thus, the displacement field is

$$\begin{Bmatrix} u \\ v \\ w \end{Bmatrix}=\sum_{i=1}^{4} N_i\left[\begin{Bmatrix} u_i \\ v_i \\ w_i \end{Bmatrix}+\frac{\zeta}{2}h_i\begin{bmatrix} l_{1i} & -l_{2i} & l_{3i} \\ m_{1i} & -m_{2i} & m_{3i} \\ n_{1i} & -n_{2i} & n_{3i} \end{bmatrix}\begin{Bmatrix} a_{xi} \\ a_{yi} \\ a_{zi} \end{Bmatrix}\right] \tag{1-39}$$

They can be rewritten in the form

$$\begin{Bmatrix} u \\ v \\ w \end{Bmatrix} = \sum_{i=1}^{4} N_i \left[\begin{Bmatrix} u_i \\ v_i \\ w_i \end{Bmatrix} + \zeta \, [b_i] \begin{Bmatrix} a_{xi} \\ a_{yi} \\ a_{zi} \end{Bmatrix} \right] = \sum_{i=1}^{4} [N]_i \boldsymbol{\delta}_i \tag{1-40}$$

where

$$[b_i] = \begin{bmatrix} b_{11i} & b_{12i} & b_{13i} \\ b_{21i} & b_{22i} & b_{23i} \\ b_{31i} & b_{32i} & b_{33i} \end{bmatrix} = \frac{h_i}{2} \, [\mathbf{V}_{1i} \; -\mathbf{V}_{2i} \; \mathbf{V}_{3i}] \tag{1-41}$$

$$\boldsymbol{\delta}_i = [u_i \; v_i \; w_i \; a_{xi} \; a_{yi} \; a_{zi}]^T \tag{1-42}$$

$$[N]_i = \begin{bmatrix} N_i & 0 & 0 & N_i\zeta b_{11i} & N_i\zeta b_{12i} & N_i\zeta b_{13i} \\ 0 & N_i & 0 & N_i\zeta b_{21i} & N_i\zeta b_{22i} & N_i\zeta b_{23i} \\ 0 & 0 & N_i & N_i\zeta b_{31i} & N_i\zeta b_{32i} & N_i\zeta b_{33i} \end{bmatrix} \tag{1-43}$$

The strains are

$$\boldsymbol{\epsilon} = \begin{Bmatrix} \dfrac{\partial u}{\partial x} \\ \dfrac{\partial v}{\partial y} \\ \dfrac{\partial u}{\partial y} + \dfrac{\partial v}{\partial x} \\ \dfrac{\partial w}{\partial z} \\ \dfrac{\partial v}{\partial z} + \dfrac{\partial w}{\partial y} \\ \dfrac{\partial w}{\partial x} + \dfrac{\partial u}{\partial z} \end{Bmatrix} = [B]\boldsymbol{\delta}_e = [B_1 B_2 B_3 B_4] \begin{Bmatrix} \boldsymbol{\delta}_1 \\ \boldsymbol{\delta}_2 \\ \boldsymbol{\delta}_3 \\ \boldsymbol{\delta}_4 \end{Bmatrix} \tag{1-44}$$

where vecotr δ_i is shown in equation (1-42),

$$\begin{aligned} c_{ix} &= N_{i,x}\zeta + N_i\zeta_{,x} \\ c_{iy} &= N_{i,y}\zeta + N_i\zeta_{,y} \\ c_{iz} &= N_{i,z}\zeta + N_i\zeta_{,z} \end{aligned} \tag{1-45}$$

and

$$[B_i] = \begin{bmatrix} N_{i,x} & 0 & 0 & b_{11i}c_{ix} & b_{12i}c_{ix} & b_{13i}c_{ix} \\ 0 & N_{i,y} & 0 & b_{21i}c_{iy} & b_{22i}c_{iy} & b_{23i}c_{iy} \\ N_{i,y} & N_{i,x} & 0 & b_{11i}c_{iy}+b_{21i}c_{ix} & b_{12i}c_{iy}+b_{22i}c_{ix} & b_{13i}c_{iy}+b_{23i}c_{ix} \\ 0 & 0 & N_{i,z} & b_{31i}c_{iz} & b_{32i}c_{iz} & b_{33i}c_{iz} \\ 0 & N_{i,z} & N_{i,y} & b_{21i}c_{iz}+b_{31i}c_{iy} & b_{22i}c_{iz}+b_{32i}c_{iy} & b_{23i}c_{iz}+b_{33i}c_{iy} \\ N_{i,z} & 0 & N_{i,x} & b_{31i}c_{ix}+b_{11i}c_{iz} & b_{32i}c_{ix}+b_{12i}c_{iz} & b_{33i}c_{ix}+b_{13i}c_{iz} \end{bmatrix} \tag{1-46}$$

In order to calculate $N_{i,x}$, $N_{i,y}$, $N_{i,z}$ and $\zeta_{,x}$, $\zeta_{,y}$, $\zeta_{,z}$, the following vectors are introduced:

$$\boldsymbol{S}=\begin{Bmatrix} x_{,\xi} \\ y_{,\xi} \\ z_{,\xi} \end{Bmatrix}=\sum_{i=1}^{4} N_{i,\xi}\left(\begin{Bmatrix} x_i \\ y_i \\ z_i \end{Bmatrix}+\frac{h_i}{2}\zeta \boldsymbol{V}_{3i}\right) \tag{1-47}$$

$$\boldsymbol{T}=\begin{Bmatrix} x_{,\eta} \\ y_{,\eta} \\ z_{,\eta} \end{Bmatrix}=\sum_{i=1}^{4} N_{i,\eta}\left(\begin{Bmatrix} x_i \\ y_i \\ z_i \end{Bmatrix}+\frac{h_i}{2}\zeta \boldsymbol{V}_{3i}\right) \tag{1-48}$$

$$\boldsymbol{V}=\begin{Bmatrix} x_{,\zeta} \\ y_{,\zeta} \\ z_{,\zeta} \end{Bmatrix}=\sum_{i=1}^{4} N_i \frac{h_i}{2} \boldsymbol{V}_{3i} \tag{1-49}$$

then, the Jacobian matrix is

$$[\,J\,] = [\,S \;\; T \;\; V\,]^T \tag{1 - 50}$$

$$[J]^{-1} = [\,\mathbf{T{\times}V} \quad \mathbf{V{\times}S} \quad \mathbf{S{\times}T}\,] / |J| \tag{1 - 51}$$

$$|J| = \mathbf{S{\times}T{\cdot}V} \tag{1 - 52}$$

Now

$$\begin{Bmatrix} N_{i,\xi} \\ N_{i,\eta} \\ N_{i,\zeta} \end{Bmatrix} = \begin{bmatrix} x_{,\xi} & y_{,\xi} & z_{,\xi} \\ x_{,\eta} & y_{,\eta} & z_{,\eta} \\ x_{,\zeta} & y_{,\zeta} & z_{,\zeta} \end{bmatrix} \begin{Bmatrix} N_{i,x} \\ N_{i,y} \\ N_{i,z} \end{Bmatrix} = [J] \begin{Bmatrix} N_{i,x} \\ N_{i,y} \\ N_{i,z} \end{Bmatrix} \tag{1-53}$$

and

$$\begin{Bmatrix} N_{i,x} \\ N_{i,y} \\ N_{i,z} \end{Bmatrix} = \begin{bmatrix} \xi_{,x} & \eta_{,x} & \zeta_{,x} \\ \xi_{,y} & \eta_{,y} & \zeta_{,y} \\ \xi_{,z} & \eta_{,z} & \zeta_{,z} \end{bmatrix} \begin{Bmatrix} N_{i,\xi} \\ N_{i,\eta} \\ N_{i,\zeta} \end{Bmatrix} = [J]^{-1} \begin{Bmatrix} N_{i,\xi} \\ N_{i,\eta} \\ N_{i,\zeta} \end{Bmatrix} \tag{1-54}$$

in which, $N_{i,\zeta} = 0$. So the expression (1-54) can be rewritten as

$$\begin{Bmatrix} N_{i,x} \\ N_{i,y} \\ N_{i,z} \end{Bmatrix} = [\,\mathbf{T{\times}V} \quad \mathbf{V{\times}S}\,] / |J| \begin{Bmatrix} N_{i,\xi} \\ N_{i,\eta} \end{Bmatrix} \tag{1-55}$$

and

$$\begin{Bmatrix} \zeta_{,x} \\ \zeta_{,y} \\ \zeta_{,z} \end{Bmatrix} = \frac{\mathbf{S{\times}T}}{|J|} \tag{1-56}$$

The stiffness matrix can be expressed in the form

$$[K]_e = \int_{-1}^{1}\int_{-1}^{1}\int_{-1}^{1} [B]^T [T]^T [C]\, [T]\, [B] \det[J]\, d\xi\, d\eta\, d\zeta \tag{1-57}$$

where

$$[T]=\begin{bmatrix} l_1^2 & m_1^2 & l_1m_1 & n_1^2 & m_1n_1 & n_1l_1 \\ l_2^2 & m_2^2 & l_2m_2 & n_2^2 & m_2n_2 & n_2l_2 \\ 2l_1l_2 & 2m_1m_2 & l_1m_2+l_2m_1 & 2n_1n_2 & m_1n_2+m_2n_1 & n_1l_2+n_2l_1 \\ l_3^2 & m_3^2 & l_3m_3 & n_3^2 & m_3n_3 & n_3l_3 \\ 2l_2l_3 & 2m_2m_3 & l_2m_3+l_3m_2 & 2n_2n_3 & m_2n_3+m_3n_2 & n_2l_3+n_3l_2 \\ 2l_3l_1 & 2m_3m_1 & l_3m_1+l_1m_3 & 2n_3n_1 & m_3n_1+m_1n_3 & n_3l_1+n_1l_3 \end{bmatrix} \tag{1-58}$$

[T] is the transformation matrix for the derivatives of displacements from global co-ordinate (x, y, z) to local co-ordinate. The direction cosines of the local co-ordinates are

$$\boldsymbol{V}_3=\begin{Bmatrix} l_3 \\ m_3 \\ n_3 \end{Bmatrix}=\frac{\sum_{i=1}^{4} N_i \boldsymbol{V}_{3i}}{\left|\sum_{i=1}^{4} N_i \boldsymbol{V}_{3i}\right|} \tag{1-59}$$

$$\boldsymbol{V}_1=\begin{Bmatrix} l_1 \\ m_1 \\ n_1 \end{Bmatrix}=\frac{\boldsymbol{i}\times\boldsymbol{V}_3}{|\boldsymbol{i}\times\boldsymbol{V}_3|} \qquad \boldsymbol{V}_2=\begin{Bmatrix} l_2 \\ m_2 \\ n_2 \end{Bmatrix}=\boldsymbol{V}_3\times\boldsymbol{V}_1 \tag{1-60}$$

This element is known to have 'locking' problems due to inconsistencies in the modelling of transverse shear energy and membrane energy [1.26-1.27]. The locking can be avoided by using reduced integration [1.26]. Although the reduced integration solution is the most economical way, the process allows some elements to exhibit spurious displacement modes. In the element proposed by Barboni and Gaudenzi [1.28], the displacement components of a higher-order element are expanded in power series along the thickness direction. This element is found to be less sensitive to locking. In general, the higher-order elements [1.29-1.30] are less prone to membrane and shear locking problems. However, the transverse strains will be continuous across the interlaminar surface in the higher-order elements. Thus, discontinuous transverse stresses will be obtained after multiplication of discontinuous material properties. This is the main drawback of the general higher-order plate theories for analysis of composite structures.

Laminated Solid Element

A special 3-D, 20-node laminated element (see in Figure 11) was developed by Hoa et al [1.31-1.34] based on the equivalent single-layer theory [1.3]. Each element contains all of the layers in the thickness of the composite structure which is assumed to be in a three dimensional stress state. In the element, the displacement field is assumed over the entire thickness,

$$\mathbf{u} = [N(\xi,\eta,\zeta)]\,\boldsymbol{\delta}_e \tag{1-61}$$

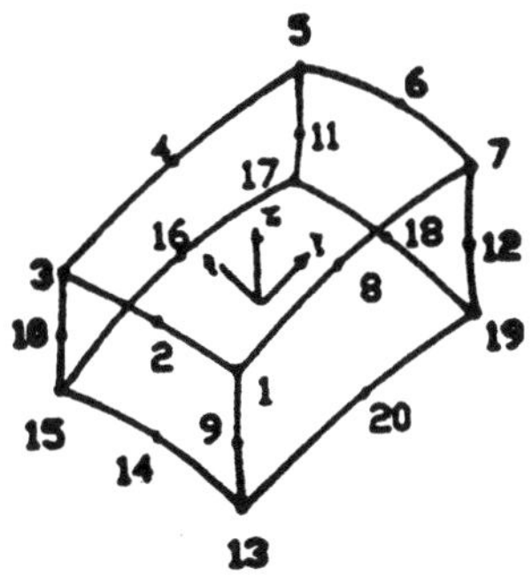

Figure 11 20-node finite element

where

$$\delta_e = [\ \delta_1\ \delta_2\ \ldots\ldots\ \delta_{20}\ ]^T \tag{1 - 62}$$

$$\delta_i = [\ u_i\ v_i\ w_i\ ]^T \quad \text{and} \quad \mathbf{u} = [\ u\ v\ w\ ]^T \tag{1 - 63}$$

$$[N] = [\ [N]_1\ [N]_2\ \ldots\ldots\ [N]_{20}\ ] \tag{1 - 64}$$

$$[N]_i = \begin{bmatrix} N_i & & \\ & N_i & \\ & & N_i \end{bmatrix} \tag{1-65}$$

With the origin of the co-ordinates at the centroid as in Figure 11, the N_i is expressed in the equation (1-26). The linear strain-displacement equations are

$$\varepsilon_x=\frac{\partial u}{\partial x} \qquad \gamma_{xy}=\frac{\partial u}{\partial y}+\frac{\partial v}{\partial x}$$
$$\varepsilon_y=\frac{\partial v}{\partial y} \qquad \gamma_{yz}=\frac{\partial v}{\partial z}+\frac{\partial w}{\partial y} \tag{1-66}$$
$$\varepsilon_z=\frac{\partial w}{\partial z} \qquad \gamma_{xz}=\frac{\partial w}{\partial x}+\frac{\partial u}{\partial z}$$

Equation (1-66) can be put in the form:

$$[\varepsilon_x \ \varepsilon_y \ \varepsilon_z \ \gamma_{yz} \ \gamma_{xz} \ \gamma_{xy}]^T=[BG]\,[\delta_1 \ \delta_2 \ \ldots\ldots\ \delta_{20}]^T \tag{1-67}$$

In the local co-ordinate $\xi\eta\zeta$ system, the elastic constant matrix [C'] can be obtained from the matrix [C] which is in the material co-ordinate system. For ease in integration and assembly, it is necessary to express the local co-ordinate strains ε' in terms of the global nodal displacements δ. They are obtained from

$$\begin{bmatrix} \frac{\partial u'}{\partial x'} & \frac{\partial v'}{\partial x'} & \frac{\partial w'}{\partial x'} \\ \frac{\partial u'}{\partial y'} & \frac{\partial v'}{\partial y'} & \frac{\partial w'}{\partial y'} \\ \frac{\partial u'}{\partial z'} & \frac{\partial v'}{\partial z'} & \frac{\partial w'}{\partial z'} \end{bmatrix} = [L]^T \begin{bmatrix} \frac{\partial u}{\partial x} & \frac{\partial v}{\partial x} & \frac{\partial w}{\partial x} \\ \frac{\partial u}{\partial y} & \frac{\partial v}{\partial y} & \frac{\partial w}{\partial y} \\ \frac{\partial u}{\partial z} & \frac{\partial v}{\partial z} & \frac{\partial w}{\partial z} \end{bmatrix} [L] \tag{1-68}$$

where

$$[L] = [\,\mathbf{V}_1\ \mathbf{V}_2\ \mathbf{V}_3\,] \tag{1 - 69}$$

$\mathbf{V}_1$, $\mathbf{V}_2$ and $\mathbf{V}_3$ are three unit normal vectors at a point P based on the direction ξ as a reference. The local co-ordinate strains can be written as

$$\varepsilon'_x=\frac{\partial u'}{\partial x'} \qquad \gamma'_{xy}=\frac{\partial u'}{\partial y'}+\frac{\partial v'}{\partial x'}$$
$$\varepsilon'_y=\frac{\partial v'}{\partial y'} \qquad \gamma'_{yz}=\frac{\partial v'}{\partial z'}+\frac{\partial w'}{\partial y'} \tag{1-70}$$
$$\varepsilon'_z=\frac{\partial w'}{\partial z'} \qquad \gamma'_{xz}=\frac{\partial w'}{\partial x'}+\frac{\partial u'}{\partial z'}$$

Above equation can be put in matrix notation as,

$$\varepsilon'_{6x1} = [BF]_{6x60}\,\delta_e \tag{1-71}$$

Now the element stiffness matrix can be evaluated as

$$[K]_e = \int_{-1}^{+1}\int_{-1}^{+1}\int_{-1}^{+1} [BF]^T [C'] \, [BF] \det[J] \, d\xi \, d\eta \, d\zeta \tag{1-72}$$

where [C'] is the material constant matrix in the local co-ordinate system. The integral is evaluated by Gauss quadrature numerical integration. However, the elasticity matrix [C'] is different from layer to layer and is not a continuous function of ζ. The thickness concept is utilized in defining the elastic properties of an individual layer to obtain the stiffness coefficients for the entire element [1.32]. This can been achieved by splitting the integration limits through each layer. The change of variable ζ is

$$\zeta = -1 + \frac{1}{t}\left[-h_k(1-\zeta_k) + 2\sum_{j=1}^{k} h_j\right] \tag{1-73}$$

and

$$d\zeta = \left(\frac{h_k}{t}\right) d\zeta_k \tag{1-74}$$

where t = the overall thickness of composite structure, and h_j = the thickness of the j-th layer. Thus, the ζ_k varies from -1 to +1 in any k-th layer and the known coefficients of the Gaussian quadrature formula can be applied. The element stiffness matrix takes the form

$$[K]_e = \sum_{k=1}^{n} \int_{-1}^{+1}\int_{-1}^{+1}\int_{-1}^{+1} [BF]^T [C'] \, [BF] \det[J] \, \frac{h_k}{t} \, d\xi \, d\eta \, d\zeta_k \tag{1-75}$$

where n is the total number of layers within the composite structure.

3-D Solid Element

If the displacement field (eq. 1-61) is defined within a layer of laminated composites, the laminated element formulation above becomes a 3-D solid element.

The element stiffness matrix is expressed in the form,

$$[K]_e=\int_{-1}^{+1}\int_{-1}^{+1}\int_{-1}^{+1}[BF]^T[C'][BF]\det[J]\,d\xi\,d\eta\,d\zeta \tag{1-76}$$

Usually 3-D solid elements are used to investigate the local effects in composite structures [1.9-1.10, 1.35-1.39]. One of the earliest attempts to use a 3-D solid element for the analysis of laminates appeared in [1.9]. For the analysis of the free-edge effect, Lucking and Hoa [1.40-1.41] used 3-D, 20-node solid elements to analyze the cross-ply laminate with a circular hole. Later investigations showed that many elements are required through the thickness of one layer to obtain accurate results. Barker et al [1.35-1.36] concluded that three linear elements were sufficient when the free-edge effect is studied. Other researchers found that two, or even three 20-node brick elements were required through the thickness of one layer in order to obtain accurate results[1.42].

Multilayer Element

Multilayer elements [1.43-1.50] can be derived based on 3-D continuum theories. Robbins and Reddy [1.51] proposed multilayer elements with separable interpolation functions. In these elements, it is assumed that the displacements, material properties and element geometry can be approximated by a sum of conveniently separable interpolation functions (i.e. each individual 3-D interpolation function can be written as the product of a 2-D interpolation function and a 1-D interpolation function). The transverse strains are assumed as a piecewise continuous distribution through the laminate thickness. The displacement field is expressed in the form,

$$\begin{aligned} u(x,y,z)&=\sum_{j=1}^{n}U_j(x,y)H_j(z)\\ v(x,y,z)&=\sum_{j=1}^{n}V_j(x,y)H_j(z)\\ w(x,y,z)&=\sum_{j=1}^{n}W_j(x,y)H_j(z)\end{aligned} \tag{1-77}$$

where (U_j , V_j , W_j) denote the nodal values of (u, v, w), n is the number of nodes through thickness and H_j are the global interpolation functions for the discretization of the displacements through thickness. For quadratic variation through each numerical layer (shown in figure 12), the number of subdivisions through thickness will be equal to $N_e = (n-1)/2$ and the functions $H_j(z)$ are given below:

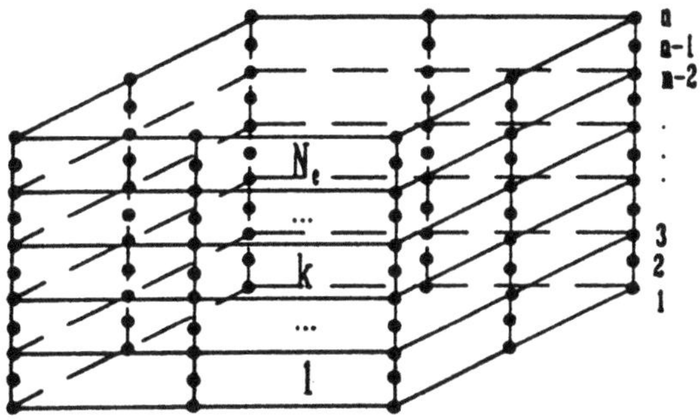

Figure 12 Multilayer element with quadratic variation through each numerical layer

$$
\begin{aligned}
H_{2k-1}(z) &= G_1^{(k)}(z) \\
H_{2k}(z) &= G_2^{(k)}(z) \qquad z_{2k-1} \le z \le z_{2k+1} \\
H_{2k+1}(z) &= G_3^{(k)}(z)
\end{aligned}
\tag{1-78}
$$

and

$$
\begin{aligned}
G_1^{(k)} &= \left(1-\frac{\bar{z}}{h_k}\right)\left(1-\frac{2\bar{z}}{h_k}\right) = -\zeta(1-\zeta)/2 \\
G_2^{(k)} &= 4\frac{\bar{z}}{h_k}\left(1-\frac{\bar{z}}{h_k}\right) = (1+\zeta)(1-\zeta) \\
G_3^{(k)} &= -\frac{\bar{z}}{h_k}\left(1-\frac{2\bar{z}}{h_k}\right) = \zeta(1+\zeta)/2
\end{aligned}
\tag{1-79}
$$

where ($k = 1, 2, 3, \ldots, N_e$), $N_e = (n-1)/2$, h_k is the thickness of the k-th layer, z =

$z - z_b^k$, and z_b^k denotes the z-coordinate of bottom of the k-th numerical layer. Furthermore, the displacement can be written as follows,

$$
\begin{aligned}
u&=\sum_{j=1}^{n}\sum_{i=1}^{8} U_{ij}\, N_i(\xi,\eta)\, H_j(\zeta)\\
v&=\sum_{j=1}^{n}\sum_{i=1}^{8} V_{ij}\, N_i(\xi,\eta)\, H_j(\zeta)\\
w&=\sum_{j=1}^{n}\sum_{i=1}^{8} W_{ij}\, N_i(\xi,\eta)\, H_j(\zeta)
\end{aligned}
\tag{1-80}
$$

where N_i is the shape function and its expression is

$$
\begin{aligned}
N_i=&\frac{1}{4}(1+\xi_0)(1+\eta_0)(\xi_0+\eta_0-1)\xi_i^2\eta_i^2\\
&+\frac{1}{2}(1-\xi^2)(1+\eta_0)(1-\xi_i^2)\eta_i^2\\
&+\frac{1}{2}(1-\eta^2)(1+\xi_0)(1-\eta_i^2)\xi_i^2
\end{aligned}
\tag{1-81}
$$

in which,

$$\xi_0=\xi_i\xi \qquad \eta_0=\eta_i\eta \qquad (i=1,2,\ldots,8) \tag{1-82}$$

where ξ_i and η_i are the local co-ordinate of node i in the element parametric space. The global co-ordinates (x,y,z) of any point within the element can be written to interpolate the local co-ordinates (ξ, η, ζ) in order to map the element geometric shape,

$$
\begin{aligned}
x&=\sum_{i=1}^{8} x_i\, N_i(\xi,\eta)\\
y&=\sum_{i=1}^{8} y_i\, N_i(\xi,\eta)\\
z&=\sum_{j=1}^{n} z_j\, H_j(\zeta)
\end{aligned}
\tag{1-83}
$$

The linear strains associated with the displacement field above are

$$\varepsilon_{xx}=\sum_{j=1}^{n}\sum_{i=1}^{8} U_{ij}\ N_{i,x}H_j$$

$$\varepsilon_{yy}=\sum_{j=1}^{n}\sum_{i=1}^{8} V_{ij}\ N_{i,y}H_j \tag{1-84}$$

$$\varepsilon_{zz}=\sum_{j=1}^{n}\sum_{i=1}^{8} W_{ij}\ N_i\ H_{j,z}$$

$$\gamma_{xy}=\sum_{j=1}^{n}\sum_{i=1}^{8} (U_{ij}\ N_{i,y}\ H_j+V_{ij}\ N_{i,x}\ H_j)$$

$$\gamma_{yz}=\sum_{j=1}^{n}\sum_{i=1}^{8} (V_{ij}\ N_i\ H_{j,z}+W_{ij}\ N_{i,y}\ H_j) \tag{1-85}$$

$$\gamma_{xz}=\sum_{j=1}^{n}\sum_{i=1}^{8} (U_{ij}\ N_i\ H_{j,z}+W_{ij}\ N_{i,x}\ H_j)$$

Note that the strains are generally discontinuous at the layer interfaces because of the layerwise definition of the functions H_j . The strain matrix is

$$\{\varepsilon\}=\begin{Bmatrix} \frac{\partial u}{\partial x} \\ \frac{\partial v}{\partial y} \\ \frac{\partial u}{\partial y}+\frac{\partial v}{\partial x} \\ \frac{\partial w}{\partial z} \\ \frac{\partial v}{\partial z}+\frac{\partial w}{\partial y} \\ \frac{\partial w}{\partial x}+\frac{\partial u}{\partial z} \end{Bmatrix}=[B]\{\mathbf{\Delta}\}^e=[B_{i1}\ B_{i2}\cdot\cdot\cdot\cdot\cdot B_{in}]\begin{Bmatrix} \mathbf{\Delta}_{i1} \\ \mathbf{\Delta}_{i2} \\ \cdot \\ \cdot \\ \mathbf{\Delta}_{in} \end{Bmatrix} \tag{1-86}$$

in which,

$$[B_{ij}]\boldsymbol{\Delta}_{ij}=[b_{1j}\ b_{2j}\cdots\cdots b_{8j}]\begin{Bmatrix}\boldsymbol{\delta}_{1j}\\ \boldsymbol{\delta}_{2j}\\ \cdot\\ \cdot\\ \boldsymbol{\delta}_{8j}\end{Bmatrix} \tag{1-87}$$

and

$$[b_{ij}]=\begin{bmatrix}N_{i,x}H_j & 0 & 0\\ 0 & N_{i,y}H_j & 0\\ N_{i,y}H_j & N_{i,x}H_j & 0\\ 0 & 0 & N_iH_{j,z}\\ 0 & N_iH_{j,z} & N_{i,y}H_j\\ N_iH_{j,z} & 0 & N_{i,x}H_j\end{bmatrix} \qquad \boldsymbol{\delta}_{ij}=\begin{Bmatrix}U_{ij}\\ V_{ij}\\ W_{ij}\end{Bmatrix} \tag{1-88}$$

where i=1,2,....,8 and j=1, 2, ..., n. In order to calculate $N_{i,x}$, $N_{i,y}$ and $H_{j,z}$, the following vectors are introduced:

$$\boldsymbol{S}=\begin{Bmatrix}x_{,\xi}\\ y_{,\xi}\end{Bmatrix}=\sum_{i=1}^{8} N_{i,\xi}\begin{Bmatrix}x_i\\ y_i\end{Bmatrix} \tag{1-89}$$

$$\boldsymbol{T}=\begin{Bmatrix}x_{,\eta}\\ y_{,\eta}\end{Bmatrix}=\sum_{i=1}^{8} N_{i,\eta}\begin{Bmatrix}x_i\\ y_i\end{Bmatrix} \tag{1-90}$$

then

$$|\, J\, | = |\ \mathbf{S}\times\mathbf{T}\ | \tag{1 - 91}$$

$$[J]^{-1}=\begin{bmatrix}y_{,\eta} & -y_{,\xi}\\ -x_{,\eta} & x_{,\xi}\end{bmatrix}/|J| \tag{1-92}$$

Because

$$\begin{Bmatrix} N_{i,\xi} \\ N_{i,\eta} \end{Bmatrix} = \begin{bmatrix} x_{,\xi} & y_{,\xi} \\ x_{,\eta} & y_{,\eta} \end{bmatrix} \begin{Bmatrix} N_{i,x} \\ N_{i,y} \end{Bmatrix} = [J] \begin{Bmatrix} N_{i,x} \\ N_{i,y} \end{Bmatrix} \tag{1-93}$$

and

$$\begin{Bmatrix} N_{i,x} \\ N_{i,y} \end{Bmatrix} = [J]^{-1} \begin{Bmatrix} N_{i,\xi} \\ N_{i,\eta} \end{Bmatrix} \tag{1-94}$$

in the thickness direction, one has

$$z_{,\zeta} = \sum_{j=1}^{3} H_{j,\zeta} z_j \tag{1-95}$$

$$H_{j,\zeta} = \frac{dz}{d\zeta} H_{j,z} \tag{1-96}$$

$$H_{j,z} = \left(\frac{dz}{d\zeta}\right)^{-1} H_{j,\zeta} \tag{1-97}$$

The stiffness matrix can be obtained using following formulations,

$$[K]_e = \int_{V_e} [B]^T [D] [B] \, dV \tag{1-98}$$

where

$$[B]^T [D] [B] = [B_{i1} B_{i2} \ldots B_{in}]^T [D] [B_{i1} B_{i2} \ldots B_{in}]$$

$$= \begin{bmatrix} B_{i1}^T D_1 B_{i1} & B_{i1}^T D_1 B_{i2} & B_{i1}^T D_1 B_{i3} & 0 & 0 & 0 & . \\ B_{i2}^T D_1 B_{i1} & B_{i2}^T D_1 B_{i2} & B_{i2}^T D_1 B_{i3} & 0 & 0 & 0 & . \\ B_{i3}^T D_1 B_{i1} & B_{i3}^T D_1 B_{i2} & B_{i3}^T (D_1 + D_2) B_{i3} & B_{i3}^T D_2 B_{i4} & B_{i3}^T D_2 B_{i5} & 0 & . \\ 0 & 0 & B_{i4}^T D_2 B_{i3} & B_{i4}^T D_2 B_{i4} & B_{i4}^T D_2 B_{i5} & 0 & . \\ 0 & 0 & B_{i5}^T D_2 B_{i3} & B_{i5}^T D_2 B_{i4} & B_{i5}^T (D_2 + D_3) B_{i5} & . & . \\ 0 & 0 & 0 & 0 & . & . & . \\ 0 & 0 & 0 & 0 & . & . & B_{in}^T D_{N_e} B_{in} \end{bmatrix} \tag{1-99}$$

In which, [D_k] is the material stiffness matrix of the k-th subdivision in the thickness of laminated composites.

1.5 ADVANTAGES AND DISADVANTAGES OF DISPLACEMENT FINITE ELEMENTS

The majority of finite elements used for stress analysis of laminated composites is based on the displacement formulation. This is due to the simple approach to the element formulation provided by the displacement model.

In the displacement finite element method, the displacement functions are assumed a priori. The finite element process first calculates the displacements (primary variables) at the nodes of the elements. The displacement field is then obtained from these nodal displacements. The strains and stresses (second quantities), which are more important for design purposes, are calculated by numerically differentiating the approximate solutions, thereby introducing additional errors. For stress analysis of homogeneous materials, the displacement finite element method can provide accurate results efficiently. However, for laminated composites, due to large stress gradients in the transverse directions, large computer space is usually required if displacement finite element method is used.

In developing finite elements for stress analysis of laminated composites, the main requirement is to satisfy the continuity conditions on displacements and transverse stresses at interlaminar surfaces, and traction-free condition on the upper and/or lower surfaces. The displacement element models can not satisfy these conditions well because the stresses are the second quantities which are calculated from approximate displacements by using numerical differentiation.

For laminated composites, the interface between layers are usually locations of large gradients of stresses and strains due to discontinuity in material properties as one moves from one layer to the next one. The use of displacement elements requires fine element mesh and extensive amount of computer space and time to be able to determine stresses and strains with any degree of accuracy. This is because of the fact that the convergence of displacement finite element model for problems with large gradients of stresses is slow. This excessive requirement of computer resources has been a deterrent to accurate and efficient stress calculation in laminated composites.

1.6 MOTIVATION FOR DEVELOPING HYBRID FINITE ELEMENTS

For laminated composites, the problem of delamination has been a great concern for designers and researchers from day one. Designers and stress analysts have been working on this problem for the past thirty years. Many numerical techniques have been proposed, and the majority of them has been using finite element method. However, until the present time, the problem has not been resolved satisfactorily. The main difficulty is in the efficiency in obtaining transverse stresses accurately. Without efficient means to obtain accurate transverse stresses, it is difficult to predict interlaminar failure.

The development of hybrid elements is motivated by attempts to overcome the disadvantages of displacement elements in order to provide an efficient and accurate method for stress analysis of laminated composites. The hybrid stress finite element formulation assumes the stresses as the independent variables at the outset. Therefore, the degree of accuracy of the stress is the same as the degree of accuracy of the displacement. This is due to the fact that the stresses are obtained directly from the process of minimization and without having to go through the differentiation of the displacements. This is the inherent advantage of the hybrid finite element method. Furthermore, the hybrid elements can exactly satisfy the continuity conditions on displacements and stresses at interlaminar surfaces of laminated composites and the finite element model using these elements can efficiently provide accurate transverse stresses.

REFERENCES

1.1 A.K. Noor, 'Mechanics of anisotropic plates and shells-a new look at an old subject', Computers & Structures, vol. 44, no.3, 499-514(1992).

1.2 J.N. Reddy & D.H. Robbins Jr., 'Theories and computational models for composite laminates', Appl. Mech. Rev., vol. 47, no. 6, part 1, 147-169(1994).

1.3 A.K. Noor and W.S. Burton, 'Assessment of shear deformation theories for multilayered composite plates', Appl. Mech. Rev., vol. 42, no. 1, 1-9(1989).

1.4 E. Reissner and Y. Stavsky, 'Bending and stretching of certain types of heterogeneous aeroisotropic elastic plates', J. Appl. Mech. ASME, 28, 402-408(1961).

1.5 K.H. Lo, R.M. Christensen and E.M.Wu, 'A higher-order theory of plate deformation: Part 1, Homogeneous plates; Part 2, Laminated plates', J. Appl. Mech. ASME, 44, 663-676(1977).

1.6 J.N. Reddy, 'A simple higher-order theory for laminated composite plates', J. Appl. Mech. ASME, 51,745-752(1984).

1.7 J.N. Reddy, 'A refined nonlinear theory of plates with transverse shear deformation', Int. J. Solids Struct., 20, 881-896(1984).

1.8 J.M. Whitney and C.T. Sun, 'A higher order theory for extensional motion of laminated composites', J. Sound Vib., vol. 30, 85-97(1973).

1.9 E.F. Rybicki, 'Approximate three-dimensional solution for symmetric laminates under in-plane loading', J. Compo. Mater., vol. 5, p354(1971).

1.10 L.B. Lessard, M.M. Shokrieh & A.S. Schmidt, '3-D stress analysis of composite plates with or without stress concentrations', Composites Modelling and Processing Science, III, ICCM/9, Ed. Antonio Miravete, Woodhead Publishing Limited, (1993).

1.11 R.M. Barker, F.T. Lin and J.R. Dana, '3-D finite element analysis of laminated composites', Computers & Structures, vol.2, 1013-1029(1972).

1.12 M. Epstein and H.P.Huttelmaier, 'A finite element formulation for multilayered and thick plates', Comp. Struct., 16, 645-650(1983).

1.13 S.T. Mau, P. Tong and T.H.H. Pian, 'Finite element solutions for laminated thick plates', J.

Composite Materials, vol.6, 304-311(1972).
1.14 R.L. Spilker, S.C. Chou and O. Orringer, 'Alternate hybrid-stress elements for analysis of multilayer composite plates', J. Composite Materials, vol.11, 51-70(1977).
1.15 R.L. Spilker, 'A hybrid stress finite element formulation for thick multilayer laminates', Computers & Structures, vol. 11, 507-514(1980).
1.16 R.L. Spilker, 'Hybrid-stress eight-node elements for thin and thick multilayer laminated plates', Int. J. Numer. Methods Engrg, vol.18, 801-828(1982).
1.17 W.-J. Liou & C.T. Sun, 'A three dimensional hybrid stress isoparametric element for the analysis of laminated composited plates', Computers & Structures, vol. 25, no. 2, 241-249(1987).
1.18 S.W. Tasi and H.T. Hahn, Introduction to Composite Materials, Technomic, 1980.
1.19 R.M. Jones, Mechanics of Composite Materials, McGraw-Hill, New York, 1975.
1.20 S.V. Hoa, Analysis for Design of Fiber Reinforced Plastic Vessels and Pipings, Technomic, 1991.
1.21 T.H.H. Pian, 'Variational principles for finite element methods in solid mechanics', ed. Z.-M. Zheng, Applied Mechanics, International Academic Publishers, Beijing, 34-42(1989).
1.22 O.C. Zienkiewicz, The Finite Element Method, 3rd Ed., Mcgraw-Hill, New York, 1977.
1.23 B. Irons and M. Loikkanen, 'An engineers'defence of the patch test', Int. J. Numer. Methods Eng., vol.19, 1391-1401(1983).
1.24 B.M. Irons, 'Engineering applications of numerical integration in stiffness methods', AIAA J., vol. 4, 2035-2037(1966).
1.25 R.D. Cook, 'Finite elements based on displacement fields', in Finite Element Handbook, eds., H. Kardestuncer and D.H. Norrie, McGraw-Hill, New York, 1987.
1.26 O.C. Zienkiewicz, R.L. Taylor and J.M. Too, 'Reduced integration techniques in general analysis of plates and shells', Int. J. Numer. Methods Eng., 3, 275-290(1971).
1.27 T. Belytschko, C.S. Tsay and W.K. Liu, 'A stabilization matrix for the bi-linear Mindlin plate element', Comp. Methods Appl. Mech. Eng., 29, 313-327(1981).
1.28 R. Barboni and P. Gaudenzi, 'Higher order finite element analysis of laminated composites', Proc. 7th Int. Conf. on Composite Materials, eds.,Y.-S. Wu, Z.-L. Gu, and R.-J. Wu, Academic Press, Beijing, 1989.
1.29 P.R. Heyliger and J.N.Reddy, 'A higher-order beam finite-element for bending and vibration problems', J. Sound Vib., 126,309-326(1988).
1.30 J.N. Reddy, 'A simple higher-order theory for laminated composite plates', J. Appl. Mech., vol. 51, 745-752(1984).
1.31 S.V. Hoa, C.W. Yu and T.S. Sankar, 'Analysis of filament wound vessel using finite elements', Composite Structures, vol. 3, 1-18(1985).
1.32 R. Natarajan, S.V. Hoa and T.S. Sankar, 'Stress analysis of filament wound tanks using 3-D finite elements', Int. J. Numer. Methods Engrg, vol. 23, 623-633(1986).
1.33 S.V. Hoa, B.H. Journeaux and L. Di Lalla, 'Computer aided design for composite structures', Proc. 7th Int. Conf. on Composite Materials, (eds. Y-S. Wu et al.), 383- 390(1989).
1.34 J. Daoust, Interlaminar Stresses in Tapered Laminates, Ph. D. Thesis, Concordia University, Montreal, Quebec, Canada, 1989.
1.35 R.M. Barker, F.T. Lin and J.R. Dana, 'Three-dimensional finite element analysis of laminated composites', Computers & Structures, vol. 2, 1013-1029.
1.36 R.M. Barker, J.R. Dana and C.W. Pryor, 'Stress concentrations near holes in laminates', J. Eng. Mech. Div., ASME, 477-488(1974).
1.37 D. Van Gemert and F. Norree, 'Finite element analysis of interlaminar stress singularities at a free edge in composite laminates', Computer Aided Design in Composite Material Technology, eds. C.A. Brebbia, et al., 1988.
1.38 M.S. Weinmann, G. Steinmetz, F.J. Arendts and C. Hansel, 'Numerical and experimental 3-D delamination behaviour of an anisotropic layered plate under compression loading', Composites Properties and Applications, ICCM/9, ed. A. Miravete, Woodhead Publishing Limited, 1993.
1.39 C.D. Maydom, A.C. Grag and M.L. Scott, 'Finite element analysis of terminating plies in advanced fibre composite panels', Advanced Composites '93, eds. T. Chandra and A.K. Dhingra, The Minerals, Metals and Materials Society, 1993.

1.40 W.M. Lucking, Analysis of Edge Problems in Statically-Loaded Fiber-reinforced Laminated Plates by Linear Elastic Theory, Ph. D. Thesis, Concordia University, Montreal, Quebec, Canada, 1989.
1.41 W.M. Lucking, S.V. Hoa and T.S. Sankar, 'The effect of geometry on interlaminar stresses of [0/90]s composite laminates with circular holes', J. Composite Materials, vol.17, 188-198(1984).
1.42 O. Hayden Griffin, Jr. 'The use of computers in the evaluation of three dimensional stress effects in composite materials products', Composite Material Design and Analysis, eds., W.P.de Wilde and W.R.Blain, Springer-Verlag, (1990).
1.43 S. Srinivas, 'A refined analysis of composite laminates', J. Sound & Vib., vol.30, no.4, 495-507(1973).
1.44 M. Di Sciuva, 'An improved shear-deformation theory for moderately thick multilayered anisotropic shells and plates', J. Appl. Mech., vol. 54, 589-596(1987).
1.45 D.R.J. Owen & Z.H. Li, 'A refined analysis of laminated plates by finite element displacement methods - I. fundamental and static analysis', Computers & Structures, vol.26, no. 6, 907-914(1987).
1.46 K. Bhaskar & T.K. Varadan, 'Refinement of higher-order laminated plate theories', AIAA J., vol. 27, no. 12, 1830-1832(1989).
1.47 C.-Y. Lee, D. Liu, and X.Q. Lu, 'Static and vibration analysis of laminated composite beams with an interlaminar shear stress continuity theory', Int. J. Numer. Methods Eng., vol. 33, 409-424(1992).
1.48 C.Y. Lee and D. Liu, 'An interlaminar stress continuity theory for laminated composite analysis', Computers & Structures, vol.42, no.1, 69-78(1992).
1.49 A.V.K. Murty and S. Vellaichamy, 'Finite element estimation of interlaminar stresses in laminated composite', Computer Aided Design in Composite Material Technology, eds., C.A.Brebbia, W.P.de Wilde and W.R. Blain, Springer-verlag, (1988).
1.50 S. Botello, E. Onate and J. Miquel, 'A layer wise finite element model for analysis of composite plates and shells', Composite Modelling and Processing Science, vol. III, ICCM/9, ed. A. Miravete, Woodhea Publishing Limited, 1993.
1.51 D.H. Robbins, Jr. and J.N. Reddy, 'Modelling of thick composites using a layer-wise laminate theory', Int. J. Numer. Methods Eng., vol.36, 655-677(1993).

Chapter 2

THE HYBRID FINITE ELEMENT METHOD

2.1 INTRODUCTION

In structural and solid mechanics, finite element practice was based primarily on single-field formulations of element properties [2.1-2.7] in the early 1960s. During this period, two major types of finite elements developed were the compatible element and the equilibrium element based, respectively, on the principles of minimum potential energy and complementary energy. For compatible finite element model, the assumed displacements are compatible both within the element and along the interelement boundary. For equilibrium finite element model, the stresses are equilibrating within the element and the tractions are balancing along the interelement boundary. These single-field finite element models provide the simplest approaches to the element formulation. However, in the compatible model, the primary variables (displacements) are first computed and then second quantities (stresses), which are more important for design purposes, are calculated by numerically differentiating the approximate solutions. The numerical differentiation results in additional errors. On the other hand, the equilibrium models have found limited use in general-purpose computer codes because they behave as mechanisms without a judicious choice of basis functions. Therefore, multifield finite elements have been developed to overcome the shortcomings of the single-field finite elements. In 1964, a multifield finite element was formulated first by Pian [2.8-2.9] by using Lagrange multipliers to enforce the constraint conditions along the interelement boundary.

Since 1964, a great number of multifield finite elements have been presented and a number of hybrid and mixed element models have been proposed [2.10-2.14]. Most of these are based on the character of the variational principle and the constraints used in the element development. For example, the *mixed element* was defined as the element which is based on a multifield variational functional, and the *hybrid element* was defined as the element which is based on the introduction of Lagrange multipliers to enforce the constraint conditions along the inter-element boundary. However, under this definition, these two types of element are not mutually exclusive. Therefore, the terms *hybrid* and *mixed* were to be redefined. In order to make hybrid and mixed elements mutually exclusive, later on, the definition is placed on the character of the variational principle and the nature of the resulting discrete algebraic equations. Thus, the term *mixed element* is defined as the one which is formulated by multifield variational functional and contains more than one field variable in the resulting matrix equations; the term *hybrid element* is defined as the one which is formulated by multifield variational functional, yet the resulting matrix equations consist of only the nodal values of displacements as unknown. Today, this definition has become popular and many authors use it to classify the multifield finite elements [2.15-2.18]. In this book, the term *hybrid element* is used under this definition. In governing equations, only displacements appear.

For analysis of composite structures, the majority of finite element analysis still uses single-field displacement elements. This is due to the simplicity in element formulation provided by displacement finite element model. It is also due to the availability of many commercial finite element codes. However, the single-field displacement finite elements suffer from inefficiency in analysis of composite structures. The disadvantages of the single-field displacement element have been discussed in the section 1.5, chapter 1. In order to overcome these disadvantages, as an alternate, the hybrid elements [2.19-2.22] have drawn more and more attention from engineers and designers of composite structures.

A hybrid element can be formulated by many different techniques. Although most of the successful finite elements were initially based on intuitive insight rather than rigorous variational principles, researchers are always keen on devising variational bases for the new elements. Variational bases are considered to be important not only for legitimacy but also for the confidence of the element users. Therefore, in this book, variational principles will be used to formulate hybrid finite elements. In this chapter, we will firstly introduce a few new variational functionals, and then, they will be used to formulate the hybrid finite elements.

2.2 FORMULATION OF THE VARIATIONAL FUNCTIONAL

In structural and solid mechanics, there are two basic variational principles: the principle of minimum potential energy and the principle of minimum complementary energy. They can be derived from the principle of virtual work [2.23-2.24]. As mentioned above, the compatible elements are derived from the principle of minimum potential energy, which has the equilibrium conditions and the traction boundary conditions as its Euler equations, and the equilibrium elements are derived from the principle of minimum complementary energy, which has the compatibility equations and the displacement boundary conditions as its Euler equations. Due to the fact that the two basic variational principles can be generalized by the introduction of Lagrange multipliers to impose the constraint conditions, a lot of different finite elements can be developed. In an approximate solution, a generalized variational principle corresponds to a relaxation such that the constraint conditions are satisfied only in the variational sense. Now, let us describe an elasticity problem first.

Elasticity Problem

Consider a linear elastic body under static loading. The body occupies the volume V and is bounded by the surface S, which is decomposed into S: $S_d \cup S_t$. Displacements are prescribed on S_d, whereas surface tractions are prescribed on S_t. The outward unit normal on S is denoted by **n**. The following relations between three fields: stress σ, strain ε, and displacement **u** in the volume have to be satisfied.

1. the strain-displacement equations:

$$\boldsymbol{\varepsilon} = \boldsymbol{D}\boldsymbol{u} \tag{2-1}$$

or

$$\varepsilon_{ij} = \frac{1}{2}\left(u_{i,j} + u_{j,i}\right) \qquad in\ V \tag{2-1'}$$

in which,

$$\boldsymbol{D} = \begin{bmatrix} \frac{\partial}{\partial x} & 0 & 0 \\ 0 & \frac{\partial}{\partial y} & 0 \\ 0 & 0 & \frac{\partial}{\partial z} \\ 0 & \frac{\partial}{\partial z} & \frac{\partial}{\partial y} \\ \frac{\partial}{\partial z} & 0 & \frac{\partial}{\partial x} \\ \frac{\partial}{\partial y} & \frac{\partial}{\partial x} & 0 \end{bmatrix}$$

2. the stress-strain equations (constitutive equations):

$$\boldsymbol{\sigma}=[C]\,\boldsymbol{\varepsilon} \qquad or \qquad \boldsymbol{\varepsilon}=[S]\,\boldsymbol{\sigma} \tag{2-2}$$

or

$$\sigma_{ij}=C_{ijkl}\varepsilon_{kl} \qquad in \quad V \tag{2-2)'}$$

and

$$\frac{\partial A(\boldsymbol{\varepsilon})}{\partial \varepsilon_{ij}}=\sigma_{ij} \qquad or \qquad \frac{\partial B(\boldsymbol{\sigma})}{\partial \sigma_{ij}}=\varepsilon_{ij} \tag{2-2)"}$$

in which, $A(\varepsilon)$ is the strain energy function, and $B(\sigma)$ is the complementary energy function.

3. the equilibrium equations:

$$\boldsymbol{D}^T\boldsymbol{\sigma}=\boldsymbol{F} \tag{2-3}$$

or

$$\sigma_{ij,j}+F_i=0 \qquad in \quad V \tag{2-3)'}$$

in which, **F** is the body force in V.

Moreover, there are two sets of boundary conditions for the displacement field and stress field:

4. the traction boundary conditions:

$$\boldsymbol{\sigma}\cdot\boldsymbol{n}=\boldsymbol{T}_n \qquad and \qquad \boldsymbol{T}_n=\boldsymbol{T} \tag{2-4}$$

or

$$\sigma_{ij}n_j=T_{ni} \qquad and \qquad T_{ni}=T_i \qquad on\ S_t \tag{2-4)'}$$

in which, **T** is the prescribed surface force on S_t.

5. the displacement boundary conditions:

$$\mathbf{u} = \mathbf{d} \tag{2-5}$$

or

$$u_i = d_i \quad on \quad S_d \tag{2-5}$$

in which, **d** is the prescribed displacement on S_d.

Principle of Minimum Potential Energy

In order to present the variational principle, it is assumed that the strain energy function A is a positive definite function of the strain components, and the body forces and surface forces are derivable from potential functions $\Omega(\mathbf{u})$ and $\Psi(\mathbf{u})$ such that

$$\begin{aligned} -\delta\Omega(\mathbf{u}) &= \mathbf{F}^T\delta\mathbf{u} \\ -\delta\Psi(\mathbf{u}) &= \mathbf{T}^T\delta\mathbf{u} \end{aligned} \tag{2-6}$$

Then, the principle of minimum potential energy [2.23] states

Among all the admissible displacement fields, the actual displacement field makes the total potential energy

$$\Pi_p = \int_V A(\mathbf{u})\,dV - \int_V \mathbf{F}^T\mathbf{u}\,dV - \int_{S_t} \mathbf{T}^T\mathbf{u}\,dS \tag{2-7}$$

an absolute minimum, i.e., $\delta\Pi_p = 0$.

In this principle, equations (2-1), (2-2), and (2-5) are constraint conditions satisfied a priori, whereas equations (2-3) and (2-4) are Euler equations [2.24]. For the linear elastic body, the strain energy function can be expressed as

$$A(\boldsymbol{\varepsilon}) = \frac{1}{2}\boldsymbol{\varepsilon}^T[C]\,\boldsymbol{\varepsilon} \tag{2-8}$$

It can be modified to

$$A(\mathbf{u}) = \frac{1}{2} (\mathbf{Du})^T [C] (\mathbf{Du}) \tag{2-8)'}$$

In which, [C] is the stiffness matrix of materials.

Principle of Minimum Complementary Energy

For the principle of minimum complementary energy, it is assumed that the complementary energy function B is a positive definite function of the stress components. The principle [2.23] states

Among all the admissible stress fields, the actual stress field makes the total complementary energy

$$\Pi_c = \int_V B(\boldsymbol{\sigma})\, dV - \int_{S_d} \mathbf{T}_n^T \mathbf{d}\, dS \tag{2-9}$$

an absolute minimum, i.e., $\delta\Pi_c = 0$.

In this principle, equations (2-2), (2-3), and (2-4) are constraint conditions satisfied a priori, whereas equations (2-1) and (2-5) are Euler equations [2.24]. For the linear elastic body, the complementary energy function can be expressed in the form

$$B(\mathbf{u}) = \frac{1}{2} \boldsymbol{\sigma}^T [S] \boldsymbol{\sigma} \tag{2-10}$$

where [S] is the material compliance matrix.

Generalized Variational Principles

The generalized variational principles can be derived from the principles of minimum potential energy and minimum complementary energy. A systematic procedure for the derivation is to impose the constraint conditions by introducing Lagrange multipliers in the variational expression. For example, the strain-displacement equations (2-1) and the displacement boundary conditions (2-5) are constraint conditions satisfied a priori in the principle of minimum potential energy. By means of introducing Lagrange multipliers **q** and **p** defined in V and S_d, a generalized variational principle can be derived as follows:

The actual solution can be given by the stationary conditions of the functional Π_I defined as

$$\Pi_I=\int_V [A(\boldsymbol{\varepsilon})-\boldsymbol{q}^T(\boldsymbol{\varepsilon}-\boldsymbol{D}\boldsymbol{u})-\boldsymbol{F}^T\boldsymbol{u}]\,dV \\ -\int_{S_t} \boldsymbol{T}^T\boldsymbol{u}\,dS-\int_{S_d} \boldsymbol{p}^T(\boldsymbol{u}-\boldsymbol{d})\,dS \tag{2-11}$$

in which,

$$\boldsymbol{q}^T=[q_{11}\ q_{22}\ q_{33}\ q_{23}\ q_{31}\ q_{12}] \\ \boldsymbol{p}^T=[p_1\ p_2\ p_3] \tag{2-12}$$

In the functional (2-11), the Lagrange multipliers **q** and **p** must be identified by means of the stationary conditions.

The number of the independent quantities subject to variation in the functional (2-11) are eighteen. These are six components in the strain ε, three components in the displacement **u**, six components in the Lagrange multipliers **q** and three components in the Lagrange multipliers **p**. By taking variations with respect to these quantities and applying divergence theorem, the expression (2-11) becomes

$$\delta\Pi_I=\int_V \{(\frac{\partial A}{\partial \varepsilon_{ij}}-q_{ij})\,\delta\boldsymbol{\varepsilon}-(\boldsymbol{\varepsilon}-\boldsymbol{D}\boldsymbol{u})^T\delta\boldsymbol{q}-(\boldsymbol{D}^T\boldsymbol{q}+\boldsymbol{F})^T\delta\boldsymbol{u}\}dV \\ +\int_{S_t}[(\boldsymbol{q}\boldsymbol{n})-\boldsymbol{T}]^T\delta\boldsymbol{u}\,dS-\int_{S_d}(\boldsymbol{u}-\boldsymbol{d})^T\delta\boldsymbol{p}\,dS+\int_{S_d}[(\boldsymbol{q}\boldsymbol{n})-\boldsymbol{p}]^T\delta\boldsymbol{u}\,dS \tag{2-13}$$

in which,

$$(\frac{\partial A}{\partial \varepsilon_{ij}}-q_{ij})=[(\frac{\partial A}{\partial \varepsilon_{xx}}-q_{11})\ \ (\frac{\partial A}{\partial \varepsilon_{yy}}-q_{22})\ \ \ldots\ \ (\frac{\partial A}{\partial \varepsilon_{xy}}-q_{12})]$$

By means of the stationary conditions, the Lagrange multipliers **q** and **p** are identified as follows,

$$q_{ij}=\frac{\partial A}{\partial \varepsilon_{ij}}=\sigma_{ij}$$
$$p_i=q_{ij}n_j \tag{2-14}$$

All equations: the strain-displacement relations (2-1), the equilibrium equations (2-3), the traction boundary conditions (2-4), and the displacement boundary conditions (2-5), are the Euler equations. Substituting equations (2-14) into the generalized functional (2-11), a three-field variational principle (**u**, ε, σ) is resulted in

$$\Pi_{II}=\int_V [\frac{1}{2}\boldsymbol{\varepsilon}^T[C]\,\boldsymbol{\varepsilon}-\boldsymbol{\sigma}^T(\boldsymbol{\varepsilon}-\boldsymbol{D}\mathbf{u})-\boldsymbol{F}^T\mathbf{u}]\,dV$$
$$-\int_{S_t}\boldsymbol{T}^T\mathbf{u}\;dS-\int_{S_d}\boldsymbol{T}_n^T(\mathbf{u}-\boldsymbol{d})\,dS \tag{2-15}$$

This generalized variational principle is well known as the Hu-Washizu principle in structural and solid mechanics. In view of mathematics [2.24], the stress-strain relations are constraint conditions in this variational principle because they are used for identifying the Lagrange multipliers **q** in equations (2-14). However, in view of the fact that the stresses σ_{ij} are used as the Lagrange multipliers at beginning and they do not have to be identified, the stress-strain relations can be considered as the Euler equations of the variational principle [2.25-2.26]. Therefore, the variational principle (2-11) is a three-field variational principle (**u**, ε, **q**) with the stress-strain relations as constraint conditions satisfied a priori because of the fact that the Lagrange multipliers **q** need to be identified, whereas the variational principle (2-15) is a three-field variational principle (**u**, ε, σ) without any constraint conditions.

By means of the stress-strain relations and divergence theorem, a two-field variational principle (**u**, σ) that is well known as the Hellinger-Reissner variational principle [2.27-2.28] can be derived from the Hu-Washizu functional (2-15) through eliminating the strain variable. Its expression can be written as

$$\Pi_{III}=-\int_V [\frac{1}{2}\boldsymbol{\sigma}^T[S]\,\boldsymbol{\sigma}+(\boldsymbol{D}^T\boldsymbol{\sigma})^T\mathbf{u}+\boldsymbol{F}^T\mathbf{u}]\,dV$$
$$+\int_{S_t}(\boldsymbol{T}_n-\boldsymbol{T})^T\mathbf{u}\;dS+\int_{S_d}\boldsymbol{T}_n^T\boldsymbol{d}\,dS \tag{2-16}$$

If the equilibrium equations (2-3) and the traction boundary conditions (2-4) are satisfied, the functional (2-16) is reduced to the complementary energy functional

(2-9). Moreover, by eliminating the stress variable in the three-field variational functional (2-15), another two-field variational functional ($\mathbf{u}$, ε) is obtained as follows [2.18,2.24],

$$\Pi_{IV}=\int_V [-\frac{1}{2}\boldsymbol{\varepsilon}^T[C]\,\boldsymbol{\varepsilon}+([C]\,\boldsymbol{\varepsilon})^T(\boldsymbol{D}\mathbf{u})-\boldsymbol{F}^T\mathbf{u}]\,dV$$
$$-\int_{S_t}\boldsymbol{T}^T\mathbf{u}\;dS-\int_{S_d}\boldsymbol{T}_n^T(\mathbf{u}-\boldsymbol{d})\;dS \qquad (2\text{-}17)$$

The Hellinger-Reissner variational principle [2.27-2.28] can be also derived from the principle of minimum complementary energy if the conditions of stress equilibrium are introduced as a posteriori constraint condition using Lagrange multiplers. Based on the Hellinger-Reissner variational principle, another three-field variational principle ($\mathbf{u}$, ε, σ) can be derived by high-order Lagrange multipliers [2.29]. The resulting functional is expressed in the form

$$\Pi_V=-\int_V [\frac{1}{2}\boldsymbol{\sigma}^T[S]\,\boldsymbol{\sigma}+(\boldsymbol{D}^T\boldsymbol{\sigma})^T\mathbf{u}+\boldsymbol{F}^T\mathbf{u}+\lambda\,(A+B-\boldsymbol{\varepsilon}^T\boldsymbol{\sigma})\,]\,dV$$
$$+\int_{S_t}(\boldsymbol{T}_n-\boldsymbol{T})^T\mathbf{u}\;dS+\int_{S_d}\boldsymbol{T}_n^T\boldsymbol{d}\,dS \qquad (2\text{-}18)$$

in which, λ is a constant, A is the strain energy function, and B is the complementary energy function. Recently, Felippa [2.30] proposed the parametrized variational principles for deriving different variational principles. A review of these principles has been given in the literature [2.31].

2.3 EVOLUTION OF THE HYBRID FINITE ELEMENT METHOD

Originally, the hybrid stress finite elements were formulated based on the principle of the minimum complementary energy and the introduction of Lagrange multiplers to enforce the constraint conditions along the inter-element boundary [2.8]. In this element formulation, the assumed stress field in the element must satisfy equilibrium equations a priori. It causes difficulty to assume an optimal stress field for the hybrid elements. Later on, it was realized that the equilibrium conditions can be relaxed if the hybrid element formulation is based on the generalized variational principles such as Hellinger-Reissner variational principle and Hu-Washizu variational principle. The stress field may satisfy the equilibrium equations only in a variational sense. Thus, the stress field can be described in the isoparametric co-ordinate system of the element, which would make the element less sensitive to mesh distortion. In this book, the hybrid finite elements will be formulated using isoparametric co-ordinate system.

For a hybrid element, the displacement field has to be assumed firstly. It is usually described by the nodal displacements as follows,

$$\mathbf{u} = [N_1 \; N_2 \; \ldots \; N_n] \begin{Bmatrix} \boldsymbol{\delta}_1 \\ \boldsymbol{\delta}_2 \\ \cdots \\ \boldsymbol{\delta}_n \end{Bmatrix} = [N] \, \boldsymbol{\delta}_e \tag{2-19}$$

where N_i is the shape function and [N] is the shape function matrix; δ_e is nodal displacement vector. The discussion about the assumed displacement field in the element has been given in section 1.3, chapter 1.

In the hybrid stress formulation, then, a stress field must be assumed independently as follows,

$$\boldsymbol{\sigma} = [\,P\,] \, \boldsymbol{\beta} \tag{2-20}$$

For example, for a 2-D, 4-node plane element, one of the assumed stress fields is

$$\begin{Bmatrix} \sigma_x \\ \sigma_y \\ \sigma_{xy} \end{Bmatrix} = \begin{bmatrix} 1 & 1 & 0 & 0 & -\xi \\ 1 & -1 & 0 & -\eta & 0 \\ 0 & 0 & 1 & \xi & \eta \end{bmatrix} \begin{Bmatrix} \beta_1 \\ \beta_2 \\ \cdot \\ \cdot \\ \beta_5 \end{Bmatrix} \tag{2-21}$$

and one of the assumed stress fields for a 3-D, 8-node solid element is

$$\begin{Bmatrix} \sigma_x \\ \sigma_y \\ \sigma_z \\ \sigma_{xy} \\ \sigma_{yz} \\ \sigma_{zx} \end{Bmatrix} = \begin{bmatrix} 1 & 1 & -1 & 0 & 0 & 0 & \zeta & 0 & \eta & \zeta & 0 & \eta & 0 & 0 & 0 & \eta\zeta & 0 & 0 \\ 1 & -1 & -1 & 0 & 0 & 0 & \zeta & \xi & 0 & -\zeta & \xi & 0 & 0 & 0 & 0 & 0 & \zeta\xi & 0 \\ 1 & 0 & 2 & 0 & 0 & 0 & 0 & \xi & \eta & 0 & -\xi & -\eta & 0 & 0 & 0 & 0 & 0 & \xi\eta \\ 0 & 0 & 0 & 1 & 0 & 0 & 0 & 0 & 0 & 0 & 0 & 0 & \zeta & \zeta & \zeta & 0 & 0 & 0 \\ 0 & 0 & 0 & 0 & 1 & 0 & 0 & 0 & 0 & 0 & 0 & 0 & \xi & -\xi & -\xi & 0 & 0 & 0 \\ 0 & 0 & 0 & 0 & 0 & 1 & 0 & 0 & 0 & 0 & 0 & 0 & \eta & 0 & -2\eta & 0 & 0 & 0 \end{bmatrix} \begin{Bmatrix} \beta_1 \\ \beta_2 \\ \cdot \\ \cdot \\ \beta_{18} \end{Bmatrix} \tag{2-22}$$

This can be expressed in the form

$$\boldsymbol{\sigma} = [\ \boldsymbol{\sigma}_1\ \boldsymbol{\sigma}_2\ \ldots\ \boldsymbol{\sigma}_m\ ] \begin{Bmatrix} \beta_1 \\ \beta_2 \\ \cdot \\ \cdot \\ \beta_m \end{Bmatrix} = [P]\,\boldsymbol{\beta} \tag{2-23}$$

In which, vectors σ_i are stress modes which are functions of the isoparametric co-ordinates, the parameters β_i are the corresponding stress parameters, and [P] is the stress matrix.

In the hybrid stress/strain formulation, furthermore, a strain field is also assumed independently. The assumed strain field can be expressed as follows,

$$\boldsymbol{\varepsilon} = [\ \boldsymbol{\varepsilon}_1\ \boldsymbol{\varepsilon}_2\ \ldots\ \boldsymbol{\varepsilon}_m\ ] \begin{Bmatrix} \alpha_1 \\ \alpha_2 \\ \cdot \\ \cdot \\ \alpha_m \end{Bmatrix} = [Q]\,\boldsymbol{\alpha} \tag{2-24}$$

in which, vectors ε_i are strain modes which are functions of the isoparametric co-ordinates, the parameters α_i are the corresponding strain parameters, and [Q] is the strain matrix. Thus, various hybrid finite elements can be formulated using the generalized variational principles.

Hybrid Stress Element

The Hellinger-Reissner variational principle contains two fields: displacement field and stress field. Satisfying the displacement boundary conditions (2-5) a priori, the variational functional (2-16) can be modified as follows,

$$\Pi_{III} = \int_V [-\frac{1}{2}\boldsymbol{\sigma}^T[S]\,\boldsymbol{\sigma} + \boldsymbol{\sigma}^T(\boldsymbol{D}\mathbf{u}) - \boldsymbol{F}^T\mathbf{u}]\,dV - \int_{S_t} \boldsymbol{T}^T\mathbf{u}\,dS \tag{2-25}$$

Within the element, the assumed displacement field and the assumed stress field have been given in equation (2-19) and (2-20). Thus,

$$\mathbf{D}\mathbf{u} = [B]\,\boldsymbol{\delta}_e \tag{2-26}$$

where [B] is the geometry matrix. Substituting equations (2-19), (2-20), and (2-26) into the functional (2-25), it is transformed to

$$\begin{aligned}\Pi_{III} = &-\frac{1}{2}\boldsymbol{\beta}^T\left(\int_V [P]^T[S]\,[P]\,dV\right)\boldsymbol{\beta} + \boldsymbol{\beta}^T\left(\int_V [P]^T[B]\,dV\right)\boldsymbol{\delta}_e \\ &-\boldsymbol{\delta}_e^T\left(\int_V [N]^T\mathbf{F}\,dV + \int_{S_t} [N]^T\mathbf{T}\,dS\right)\end{aligned} \tag{2-27}$$

Denote

$$\begin{aligned}[H] &= \int_V [P]^T[S]\,[P]\,dV \\ [G] &= \int_V [P]^T[B]\,dV \\ \mathbf{f}_e &= \int_V [N]^T\mathbf{F}\,dV + \int_{S_t} [N]^T\mathbf{T}\,dS\end{aligned} \tag{2-28}$$

where [H] is the flexibility matrix, [G] is the leverage matrix, and $\mathbf{f}_e$ is the equivalent nodal force vector. Thus, the functional (2-27) can be rewritten in the form,

$$\Pi_{III} = -\frac{1}{2}\boldsymbol{\beta}^T[H]\,\boldsymbol{\beta} + \boldsymbol{\beta}^T[G]\,\boldsymbol{\delta}_e - \boldsymbol{\delta}_e^T\mathbf{f}_e \tag{2-29}$$

In this variational functional, there are two independent variables β and δ_e subject to variation. From the partial stationary condition with respect to β,

$$\frac{\partial \Pi_{III}}{\partial \boldsymbol{\beta}} = 0 \tag{2-30}$$

the relation between stress parameters β and nodal displacements δ_e is obtained,

$$[H]\,\boldsymbol{\beta}=[G]\,\boldsymbol{\delta}_e \tag{2-31}$$

By means of this relation, then, the functional (2-29) becomes

$$\Pi_{III}=\frac{1}{2}\boldsymbol{\delta}_e^T\left([G]^T[H]^{-1}[G]\right)\boldsymbol{\delta}_e-\boldsymbol{\delta}_e^T\boldsymbol{f}_e \tag{2-32}$$

It can be rewritten as

$$\Pi_{III}=\frac{1}{2}\boldsymbol{\delta}_e^T[K]\,\boldsymbol{\delta}_e-\boldsymbol{\delta}_e^T\boldsymbol{f}_e \tag{2-33}$$

in which, [K] is the element stiffness matrix. It can be expressed in the form,

$$[K]=[G]^T[H]^{-1}[G] \tag{2-34}$$

From the partial stationary condition with respect to $\boldsymbol{\delta}_e$, the governing equation of the element is obtained,

$$[K]\;\boldsymbol{\delta}_e=\boldsymbol{f}_e \tag{2-35}$$

When the element equations are obtained, the global equations of the hybrid finite element model for analysis of structures can be established. The procedure is the same as that in the single-field displacement finite element model discussed in the section 1.2, chapter 1.

Hybrid Strain Element

The two-field variational functional (2-17) is different from the Hellinger-Reissner variational principle. The two fields in the functional are displacement field and strain field. Satisfying the displacement boundary conditions (2-5) a priori, the variational functional (2-17) can be modified to the form,

$$\Pi_{IV}=\int_V [-\frac{1}{2}\boldsymbol{\varepsilon}^T[C]\boldsymbol{\varepsilon}+([C]\boldsymbol{\varepsilon})^T(\boldsymbol{D}\mathbf{u})-\boldsymbol{F}^T\mathbf{u}]\,dV-\int_{S_t}\boldsymbol{T}^T\mathbf{u}\,dS \tag{2-36}$$

Within the element, the assumed displacement field and the assumed strain field have been given in equation (2-19) and (2-24). Thus,

$$\boldsymbol{D}\mathbf{u}=[B]\boldsymbol{\delta}_e \tag{2-26}$$

Substituting equations (2-19), (2-24), and (2-26) into the functional (2-36), it is transformed to

$$\begin{aligned}\Pi_{IV}=&-\frac{1}{2}\boldsymbol{\alpha}^T(\int_V[Q]^T[C][Q]\,dV)\boldsymbol{\alpha}+\boldsymbol{\alpha}^T(\int_V[Q]^T[C][B]\,dV)\boldsymbol{\delta}_e\\&-\boldsymbol{\delta}_e^T(\int_V[N]^T\boldsymbol{F}\,dV+\int_{S_t}[N]^T\boldsymbol{T}\,dS)\end{aligned} \tag{2-37}$$

Denote

$$\begin{aligned}[L]&=\int_V[Q]^T[C][Q]\,dV\\ [M]&=\int_V[Q]^T[C][B]\,dV\\ \boldsymbol{f}_e&=\int_V[N]^T\boldsymbol{F}\,dV+\int_{S_t}[N]^T\boldsymbol{T}\,dS\end{aligned} \tag{2-38}$$

Thus, the functional (2-37) can be rewritten in the form,

$$\Pi_{IV}=-\frac{1}{2}\boldsymbol{\alpha}^T[L]\boldsymbol{\alpha}+\boldsymbol{\alpha}^T[M]\boldsymbol{\delta}_e-\boldsymbol{\delta}_e^T\boldsymbol{f}_e \tag{2-39}$$

In this functional, there are two independent variables subject to variation. This is the same as that in hybrid stress element. From the partial stationary condition with respect to α,

$$\frac{\partial \Pi_{IV}}{\partial \boldsymbol{\alpha}}=0 \tag{2-40}$$

the relation between strain parameters α and nodal displacements δ_e is obtained,

$$[L]\,\boldsymbol{\alpha}=[M]\,\boldsymbol{\delta}_e \tag{2-41}$$

Using this relation, the functional (2-39) becomes

$$\Pi_{IV}=\frac{1}{2}\boldsymbol{\delta}_e^T(\,[M]^T[L]^{-1}[M]\,)\,\boldsymbol{\delta}_e-\boldsymbol{\delta}_e^T\boldsymbol{f}_e \tag{2-42}$$

It can be rewritten as

$$\Pi_{IV}=\frac{1}{2}\boldsymbol{\delta}_e^T[K]\,\boldsymbol{\delta}_e-\boldsymbol{\delta}_e^T\boldsymbol{f}_e \tag{2-43}$$

in which, [K] is the element stiffness matrix and it can be expressed in the form [2.32],

$$[K]=[M]^T[L]^{-1}[M] \tag{2-44}$$

From the partial stationary condition with respect to δ_e, the governing equation of the element is obtained,

$$[K]\;\boldsymbol{\delta}_e=\boldsymbol{f}_e \tag{2-45}$$

Hybrid Stress / Strain Element - 1

By means of generalized variational principles, we can formulate not only two-field hybrid elements, but also three-field hybrid elements. The Hu-Washizu variational principle is a three-field generalized variational principle. It contains three fields: displacement field, stress field, and strain field. Satisfying the displacement boundary conditions (2-5) a priori, the variational functional (2-15) can be modified as follows,

$$\Pi_{II}=\int_V [\frac{1}{2}\boldsymbol{\varepsilon}^T[C]\boldsymbol{\varepsilon}-\boldsymbol{\sigma}^T(\boldsymbol{\varepsilon}-\boldsymbol{D}\mathbf{u})-\boldsymbol{F}^T\mathbf{u}]\,dV-\int_{S_t}\boldsymbol{T}^T\mathbf{u}\,dS \tag{2-46}$$

Within the element, the assumed displacement field, the assumed stress field, and assumed strain field have been given in equations (2-19), (2-20), and (2-24). Thus,

$$\boldsymbol{D}\mathbf{u}=[B]\,\boldsymbol{\delta}_e \tag{2-26}$$

Substituting equations (2-19), (2-20), (2-24) and (2-26) into the functional (2-46), it is transformed to

$$\begin{aligned}\Pi_{II}=&\frac{1}{2}\boldsymbol{\alpha}^T(\int_V [Q]^T[C]\,[Q]\,dV)\,\boldsymbol{\alpha}-\boldsymbol{\alpha}^T(\int_V [Q]^T[P]\,dV)\,\boldsymbol{\beta}\\&+\boldsymbol{\beta}^T(\int_V [P]^T[B]\,dV)\,\boldsymbol{\delta}_e-\boldsymbol{\delta}_e^T(\int_V [N]^T\boldsymbol{F}\,dV+\int_{S_t}[N]^T\boldsymbol{T}\,dS)\end{aligned} \tag{2-47}$$

Denote

$$\begin{aligned}[L]&=\int_V [Q]^T[C]\,[Q]\,dV\\ [W]&=\int_V [Q]^T[P]\,dV\\ [G]&=\int_V [P]^T[B]\,dV\\ \boldsymbol{f}_e&=\int_V [N]^T\boldsymbol{F}\,dV+\int_{S_t}[N]^T\boldsymbol{T}\,dS\end{aligned} \tag{2-48}$$

Thus, the functional (2-47) can be rewritten in the form,

$$\Pi_{II}=\frac{1}{2}\boldsymbol{\alpha}^T[L]\,\boldsymbol{\alpha}-\boldsymbol{\alpha}^T[W]\,\boldsymbol{\beta}+\boldsymbol{\beta}^T[G]\,\boldsymbol{\delta}_e-\boldsymbol{\delta}_e^T\boldsymbol{f}_e \tag{2-49}$$

In this variational, there are three independent sets of variables (α, β and δ_e) subject to variation. From the partial stationary condition with respect to α,

$$\frac{\partial\Pi_{II}}{\partial\boldsymbol{\alpha}}=0 \tag{2-50}$$

the relation between strain parameters α and stress parameters β is obtained as follows,

$$[L]\,\boldsymbol{\alpha}=[W]\,\boldsymbol{\beta} \tag{2-51}$$

Using this relation, the functional (2-49) becomes

$$\Pi_{II}=-\frac{1}{2}\boldsymbol{\beta}^T([W]^T[L]^{-1}[W])\,\boldsymbol{\beta}+\boldsymbol{\beta}^T[G]\,\boldsymbol{\delta}_e-\boldsymbol{\delta}_e^T\boldsymbol{f}_e \tag{2-52}$$

It can be rewritten as

$$\Pi_{II}=-\frac{1}{2}\boldsymbol{\beta}^T[M]\,\boldsymbol{\beta}+\boldsymbol{\beta}^T[G]\,\boldsymbol{\delta}_e-\boldsymbol{\delta}_e^T\boldsymbol{f}_e \tag{2-53}$$

In which,

$$[M]=[W]^T[L]^{-1}[W] \tag{2-54}$$

From the partial stationary condition with respect to β,

$$\frac{\partial\Pi_{II}}{\partial\boldsymbol{\beta}}=0 \tag{2-55}$$

the relation between stress parameters β and nodal displacements δ_e are obtained,

$$[M]\,\boldsymbol{\beta} = [G]\,\boldsymbol{\delta}_e \tag{2-56}$$

Using this relation, the functional (2-53) becomes

$$\Pi_{II} = \frac{1}{2}\boldsymbol{\delta}_e^T\left([G]^T[M]^{-1}[G]\right)\boldsymbol{\delta}_e - \boldsymbol{\delta}_e^T\boldsymbol{f}_e \tag{2-57}$$

It can be rewritten as

$$\Pi_{II} = \frac{1}{2}\boldsymbol{\delta}_e^T[K]\,\boldsymbol{\delta}_e - \boldsymbol{\delta}_e^T\boldsymbol{f}_e \tag{2-58}$$

where [K] is the element stiffness matrix. It can be expressed in the form,

$$[K] = [G]^T[M]^{-1}[G] \tag{2-59}$$

From the partial stationary condition with respect to δ_e, the governing equation of the element is obtained,

$$[K]\;\boldsymbol{\delta}_e = \boldsymbol{f}_e \tag{2-60}$$

Substituting equation (2-54) into the equation (2-59), the element stiffness matrix can be expressed in the form [2.32-2.33],

$$[K] = [G]^T[W]^{-1}[L]\;[W]^{-1}[G] \tag{2-61}$$

Hybrid Stress / Strain Element - 2

Here, we present a general hybrid element formulation using another variational functional proposed by Chien [2.24]. It also contains three fields:

displacement field, stress field, and strain field. Satisfying the displacement boundary conditions (2-5) a priori, the variational functional (2.18) is expressed in the form,

$$\Pi_V=\int_V\left[\frac{1}{2}\boldsymbol{\varepsilon}^T[C]\,\boldsymbol{\varepsilon}-\boldsymbol{\sigma}^T(\boldsymbol{\varepsilon}-\boldsymbol{D}\mathbf{u})-\boldsymbol{F}^T\mathbf{u}\right.$$
$$\left.-\lambda\,(A+B-\boldsymbol{\varepsilon}^T\boldsymbol{\sigma})\right]dV-\int_{S_t}\boldsymbol{T}^T\mathbf{u}\;dS \qquad (2\text{-}62)$$

in which, λ is constant, A is the strain energy function (2-8), and B is the complementary energy function (2-10). Within the element, the assumed displacement field, the assumed stress field, and assumed strain field have been given in equations (2-19), (2-20), and (2-24). Thus,

$$\boldsymbol{D}\mathbf{u}=[B]\,\boldsymbol{\delta}_e \qquad (2\text{-}26)$$

Substituting equations (2-19), (2-20); (2-24) and (2-26) into the functional (2-62), it is transformed to

$$\Pi_V=\frac{1-\lambda}{2}\boldsymbol{\alpha}^T\left(\int_V[Q]^T[C]\,[Q]\,dV\right)\boldsymbol{\alpha}-\frac{\lambda}{2}\boldsymbol{\beta}^T\left(\int_V[P]^T[S]\,[P]\,dV\right)\boldsymbol{\beta}$$
$$-(1-\lambda)\,\boldsymbol{\alpha}^T\left(\int_V[Q]^T[P]\,dV\right)\boldsymbol{\beta}+\boldsymbol{\beta}^T\left(\int_V[P]^T[B]\,dV\right)\boldsymbol{\delta}_e$$
$$-\boldsymbol{\delta}_e^T\left(\int_V[N]^T\boldsymbol{F}\,dV+\int_{S_t}[N]^T\boldsymbol{T}\,dS\right) \qquad (2\text{-}63)$$

Denote

$$
\begin{aligned}
[L] &= \int_V [Q]^T [C] [Q] \, dV \\
[H] &= \int_V [P]^T [S] [P] \, dV \\
[W] &= \int_V [Q]^T [P] \, dV \\
[G] &= \int_V [P]^T [B] \, dV \\
\mathbf{f}_e &= \int_V [N]^T \mathbf{F} \, dV + \int_{S_t} [N]^T \mathbf{T} \, dS
\end{aligned}
\tag{2-64}
$$

Thus, the functional (2-63) can be rewritten in the form,

$$
\Pi_V = \frac{1-\lambda}{2} \boldsymbol{\alpha}^T [L] \boldsymbol{\alpha} - \frac{\lambda}{2} \boldsymbol{\beta}^T [H] \boldsymbol{\beta} - (1-\lambda) \boldsymbol{\alpha}^T [W] \boldsymbol{\beta} + \boldsymbol{\beta}^T [G] \boldsymbol{\delta}_e - \boldsymbol{\delta}_e^T \mathbf{f}_e
\tag{2-65}
$$

Also, there are three independent variables subject to variation in this variational. From the partial stationary condition with respect to α,

$$
\frac{\partial \Pi_{II}}{\partial \boldsymbol{\alpha}} = 0
\tag{2-66}
$$

The relation between strain parameter α and stress parameters β are derived,

$$
[L] \boldsymbol{\alpha} = [W] \boldsymbol{\beta}
\tag{2-67}
$$

Then, the functional (2-65) becomes

$$
\begin{aligned}
\Pi_V = \frac{1}{2} \boldsymbol{\beta}^T \{ -\lambda [H] - (1-\lambda) [W]^T [L]^{-1} [W] \} \boldsymbol{\beta} \\
+ \boldsymbol{\beta}^T [G] \boldsymbol{\delta}_e - \boldsymbol{\delta}_e^T \mathbf{f}_e
\end{aligned}
\tag{2-68}
$$

It can be rewritten as

$$\Pi_V=-\frac{1}{2}\boldsymbol{\beta}^T[M]\,\boldsymbol{\beta}+\boldsymbol{\beta}^T[G]\,\boldsymbol{\delta}_e-\boldsymbol{\delta}_e^T\boldsymbol{f}_e \tag{2-69}$$

In which,

$$[M]=\lambda\,[H]+(1-\lambda)\,[W]^T[L]^{-1}[W] \tag{2-70}$$

From the partial stationary condition with respect to β,

$$\frac{\partial\Pi_V}{\partial\boldsymbol{\beta}}=0 \tag{2-71}$$

the relation between stress parameters β and nodal displacements δ_e is obtained,

$$[M]\,\boldsymbol{\beta}=[G]\,\boldsymbol{\delta}_e \tag{2-72}$$

Substituting this equation into the functional (2-69), it becomes

$$\Pi_V=\frac{1}{2}\boldsymbol{\delta}_e^T(\,[G]^T[M]^{-1}[G]\,)\,\boldsymbol{\delta}_e-\boldsymbol{\delta}_e^T\boldsymbol{f}_e \tag{2-73}$$

It can be rewritten as

$$\Pi_V=\frac{1}{2}\boldsymbol{\delta}_e^T[K]\,\boldsymbol{\delta}_e-\boldsymbol{\delta}_e^T\boldsymbol{f}_e \tag{2-74}$$

where [K] is the element stiffness matrix. It can be expressed in the form,

$$[K]=[G]^T[M]^{-1}[G] \tag{2-75}$$

From the partial stationary condition with respect to δ_e, the governing equation of the element is obtained,

$$[K]\ \boldsymbol{\delta}_e = \boldsymbol{f}_e \tag{2-76}$$

Substituting equation (2-70) into the equation (2-75), the element stiffness matrix can be expressed in the form,

$$[K] = [G]^T \{\lambda [H] + (1-\lambda)\ [W]^T [L]^{-1} [W]\}^{-1} [G] \tag{2-77}$$

This stiffness matrix formulation is a general form of the stiffness matrix for hybrid elements. It can be reduced to one (2-34) based on the Hellinger-Reissner variational principle when $\lambda=1$ and one (2-61) based on the Hu-Washizu variational principle when $\lambda=0$ [2.24,2.29]. In the formulation, the constant λ is bounded between 0 and 1 so as to ensure that the stiffness matrix [K] is semi-positive. There are some approaches such as perturbation, energy balance and locking alleviation [2.34-2.38] for determination of the constant λ. For example, the constant λ can be determined in such a way that the energy stored inside the element is close to the analytically-derived energy under a typical deformation of the element.

Although there are many different types of the hybrid finite elements such as hybrid stress, hybrid strain, and hybrid stress/strain elements, the discussion in this book will be restricted on the hybrid stress method due to the fact that the hybrid stress element has been widely used in the structural and solid mechanics. Before we present the new hybrid element techniques for stress analysis of composite structures, the assumed stress field has to be discussed further because the advantages and disadvantages of hybrid stress elements are effected by introduction of the assumed stress field.

2.4 ASSUMED STRESS FIELD

For the hybrid stress element, the physical fields that must be independently assumed within the element at the beginning are not only displacement field, but also stress field. An assumed stress field consists of a set of stress modes and a set of the corresponding stress parameters. Although a number of mathematical functions such as trigonometric series and exponential functions can be used as stress mode functions, orthogonal polynomials are more appropriate as stress mode functions due to their ease and simplification. It is similar to the basis functions in the assumed displacement field. However, while the displacement polynomial is constrained by the number of displacement nodal degrees of freedom in the element, the stress polynomials have no such constraint. If an assumed stress field does not contain enough stress modes, the rank of the element stiffness matrix will be less than the total degrees of deformation freedom and the numerical solution of the finite element model will be unstable. In that case, there may be kinematic deformation modes. It is possible to suppress kinematic deformation modes by adding stress modes of higher order term, but this can not guarantee that all kinematic deformation modes are suppressed. Moreover, each extra term will add more stiffness[2.39] and overuse of stress modes will cost more computational time because the calculation of element stiffness matrix requires inversion of the flexibility matrix. The lack of a rational way for deriving the optimal assumed stress modes has obstructed the development of the hybrid finite element method.

2.4.1 Stability Condition

Some mathematical basis for the stability of the numerical solution of the hybrid finite element model has been established and a number of approaches for obtaining the optimal stress modes have been proposed. A *necessary condition* to avoid kinematic deformation modes [2.12, 2.40] is

> *The total number of stress modes in an assumed stress field must be equal to or larger than the total number of nodal displacements minus the number of rigid body modes in an element.*

or

$$m \geq n - r \qquad (2\text{-}78)$$

in which, m is the total number of stress modes in an assumed stress field, n is the total number of nodal displacements, and r is the number of rigid body modes in an element.

Brezzi [2.41], Babuska, Oden and Lee [2.42] presented necessary and sufficient conditions for stability and convergence of a hybrid element by means of functional analysis. However this can be used only as a posteriori check on a formulation. Fraeijs de Veubeke [2.43] presented a limitation principle for hybrid elements based on the Hellinger-Reissner variational principle. This work was extended to the hybrid stress/strain elements based on the Hu-Washizu variational principle [2.44]. The limitation principle [2.43] states that a hybrid element would be equivalent to its displacement counterpart if a stress space consisted of all the displacement-derived stress modes is a subspace of the assumed stress. This means that a hybrid element would be no different to a displacement element when the assumed stress field contains all stress modes derived from the assumed displacement field. Because the displacement element is always free from any spurious kinematic deformation modes, for a hybrid element to avoid spurious kinematic deformation modes, *a sufficient condition* is that the assumed stress field contains all displacement-derived stress modes.

For 2-D plane elements and 3-D brick elements, in an assumed displacement field, the polynomial terms used for interpolation are the least-order polynomial terms and usually same for every component of displacement. For example, in the assumed displacement field of a 2-D, 4-node rectangular element, every component of displacement contains four, least-order polynomial terms (see equation (1-16)). Due to the fact that the order of the polynomial terms in displacement-derived stress field is always equal to or less than that of polynomial terms in displacement field, all stress modes derived from the displacement field will be included in an assumed stress field if the polynomial terms in every stress component of the stress field are the same as those in every displacement component. Therefore, a *sufficient condition* to avoid spurious kinematic deformation modes in 2-D and 3-D elements is

> *If every stress component of an assumed stress field contains the same polynomial terms as those in every displacement component of the displacement field, the resulting hybrid element is free from any kinematic deformation modes.*

Using this sufficient condition, an assumed stress field termed *sufficient stress field* can be established based on the assumed displacement field of a hybrid element.

2.4.2 Approaches to Obtain An Assumed Stress Field

There are a few approaches for determining an assumed stress field. Using group theory, Punch and Atluri [2.45,2.46] established a set of least-order stable invariant stress selections for three-dimensional brick elements and two-dimensional rectangular elements. Pian and Chen [2.47] used the product $\{\sigma\}^T$ $\{\varepsilon\}$, the deformation energy due to the assumed stresses and displacements, to determine the necessary assumed stress modes. Pian and Tong [2.48] introduced the internal displacement parameters to relax the stress equilibrium condition and used isoparametric interpolation to construct hybrid element. Pian and Wu [2.49-2.50] introduced incompatible displacements to maintain completeness of the polynomials. The initial choice of stress terms are unconstrained and complete polynomials. The additional displacements are used as Lagrange multipliers to enforce the stress equilibrium constraint. Chen et al [2.51-2.52] constrained the stress by setting the inner product of the non-constant stress modes with the deformation derived from the incompatible displacement to zero. Sze [2.53-2.55] used orthogonal lower- and higher-order stress modes to construct hybrid element. It allows the partition of the element stiffness matrix into a lower- and a higher-order stiffness matrix. When the lower-order stiffness turns out to be identical to the sub-integrated element, the higher-order stiffness matrix plays the role of stabilization matrix. Other methods for determining the assumed stress field are referred to the literature [2.49-2.55]. In this section, three approaches for determining the assumed stress field are presented.

a) Sufficient Stress Field

The sufficient condition above indicates that if every stress component uses the same polynomial terms which are used for every displacement component, its resulting hybrid element is free from any spurious kinematic deformatiom modes. Thus, an assumed stress field can be established through the assumed displacement field.

1. 2-D, 3-node Triangular Element

For this element, a displacement field is usually assumed in the form,

$$u = a_0 + a_1\xi + a_2\eta$$
$$v = b_0 + b_1\xi + b_2\eta \tag{2-79}$$

The polynomial terms used in every displacement component are same. They are the three, least-order polynomial terms

$$[\ 1\ \ \xi\ \ \eta\ ] \tag{2-80}$$

According to the sufficient condition, every stress component uses the same terms as follows,

$$\sigma_x = \beta_1 + \beta_4\xi + \beta_7\eta$$
$$\sigma_y = \beta_2 + \beta_5\xi + \beta_8\eta \tag{2-81}$$
$$\sigma_{xy} = \beta_3 + \beta_6\xi + \beta_9\eta$$

Using this assumed stress field, the resulting hybrid element is free from any kinematic deformation modes.

2. 2-D, 4-node Rectangular Element

For this element, the number of degrees of freedom is eight and assumed displacement field usually has eight parameters. It can be expressed in the form,

$$u = a_0 + a_1\xi + a_2\eta + a_3\xi\eta$$
$$v = b_0 + b_1\xi + b_2\eta + b_3\xi\eta \tag{2-82}$$

Every displacement component uses the same least-order polynomial terms. They are

$$[\ 1\ \ \xi\ \ \eta\ \ \xi\eta\ ] \tag{2-83}$$

According to the sufficient condition, similarly, every stress component uses the same terms as follows,

$$\begin{aligned}
\sigma_x &= \beta_1+\beta_4\xi+\beta_7\eta+\beta_{10}\xi\eta \\
\sigma_y &= \beta_2+\beta_5\xi+\beta_8\eta+\beta_{11}\xi\eta \\
\sigma_{xy} &= \beta_3+\beta_6\xi+\beta_9\eta+\beta_{12}\xi\eta
\end{aligned} \tag{2-84}$$

3. 3-D, 8-node Brick Element

For a 3-D, 8-node hybrid element, the displacement field is assumed in the form,

$$\begin{aligned}
u &= a_0+a_1\xi+a_2\eta+a_3\zeta+a_4\xi\eta+a_5\xi\zeta+a_6\eta\zeta+a_7\xi\eta\zeta \\
v &= b_0+b_1\xi+b_2\eta+b_3\zeta+b_4\xi\eta+b_5\xi\zeta+b_6\eta\zeta+b_7\xi\eta\zeta \\
v &= c_0+c_1\xi+c_2\eta+c_3\zeta+c_4\xi\eta+c_5\xi\zeta+c_6\eta\zeta+c_7\xi\eta\zeta
\end{aligned} \tag{2-85}$$

The eight, least-order polynomial terms used in every displacement component are

$$[\ 1\ \ \xi\ \ \eta\ \ \zeta\ \ \xi\eta\ \ \eta\zeta\ \ \zeta\xi\ \ \xi\eta\zeta\ ] \tag{2-86}$$

According to the sufficient condition, similarly, every stress component uses the same terms as follows,

$$\begin{aligned}
\sigma_x &= \beta_1+\beta_4\xi+\beta_7\eta+\beta_{10}\zeta+\beta_{13}\xi\eta+\beta_{16}\eta\zeta+\beta_{19}\zeta\xi+\beta_{22}\xi\eta\zeta \\
\sigma_y &= \beta_2+\beta_5\xi+\beta_8\eta+\beta_{11}\zeta+\beta_{14}\xi\eta+\beta_{17}\eta\zeta+\beta_{20}\zeta\xi+\beta_{23}\xi\eta\zeta \\
\sigma_{xy} &= \beta_3+\beta_6\xi+\beta_9\eta+\beta_{12}\zeta+\beta_{15}\xi\eta+\beta_{18}\eta\zeta+\beta_{21}\zeta\xi+\beta_{24}\xi\eta\zeta
\end{aligned} \tag{2-87}$$

Using this way to establish assumed stress field, the advantage is the ease in selection of terms, but the number of terms used in assumed stress field is large. This disadvantage will affect the efficiency of the finite element model of structures. Moreover, it can be used only for these elements whose polynomial terms used for interpolation are same for every component of displacement.

b) Equilibrating Stress Field

An assumed stress field can be derived from a complete polynomial by means of equilibrium equations. Two examples are given here.

1. 2-D, 4-node Rectangular Element

For two dimensional problems, the assumed stress field for a 4-node hybrid element is firstly expressed in complete linear terms in the co-ordinates ξ and η. It is

$$\begin{aligned} \sigma_x &= \beta_1 + \beta_2\xi + \beta_3\eta \\ \sigma_y &= \beta_4 + \beta_5\xi + \beta_6\eta \\ \tau_{xy} &= \beta_7 + \beta_8\xi + \beta_9\eta \end{aligned} \tag{2-88}$$

Substituting the stress components (2-88) into the equilibrium equations (2-3), the relations between stress parameters are obtained,

$$\beta_8 = -\beta_6 \qquad and \qquad \beta_9 = -\beta_2 \tag{2-89}$$

The assumed stress field (2-88) is modified as follows,

$$\begin{Bmatrix} \sigma_x \\ \sigma_y \\ \sigma_{xy} \end{Bmatrix} = \begin{bmatrix} 1 & \xi & \eta & 0 & 0 & 0 & 0 \\ 0 & 0 & 0 & 1 & \xi & \eta & 0 \\ 0 & -\eta & 0 & 0 & 0 & -\xi & 1 \end{bmatrix} \begin{Bmatrix} \beta_1 \\ \beta_2 \\ \cdot \\ \cdot \\ \beta_7 \end{Bmatrix} \tag{2-90}$$

This assumed stress field for the hybrid element satisfies the equilibrium condition a priori. However the resulting hybrid element may have kinematic deformation modes.

2. Plate/shell Element

For plate and shell, it is more complicated to assume an equilibrating stress field within a hybrid element. First of all, the equilibrium equations (2-3) should be modified by introducing the normalized transverse co-ordinate ζ as follows,

$$\begin{aligned} &\frac{\partial\sigma_x}{\partial x}+\frac{\partial\sigma_{xy}}{\partial y}+\frac{1}{t}\frac{\partial\sigma_{xz}}{\partial\zeta}=0 \\ &\frac{\partial\sigma_{xy}}{\partial x}+\frac{\partial\sigma_y}{\partial y}+\frac{1}{t}\frac{\partial\sigma_{yz}}{\partial\zeta}=0 \\ &\frac{\partial\sigma_{xz}}{\partial x}+\frac{\partial\sigma_{yz}}{\partial y}+\frac{1}{t}\frac{\partial\sigma_z}{\partial\zeta}=0 \end{aligned} \tag{2-91}$$

where t is the half-thickness of plate. Then, the in-plane stresses are assumed to be complete cubic polynomials and the expression of the transverse stresses are obtained by using the equilibrium equations (2-91) as follows [2.56-2.57],

$$\begin{aligned} \sigma_x=&(\beta_1+\beta_2x+\beta_3y+\beta_4x^2+\beta_5xy+\beta_6y^2+\beta_7x^3+\beta_8x^2y+\beta_9xy^2+\beta_{10}y^3) \\ &+\zeta(\beta_{31}+\beta_{32}x+\beta_{33}y+\beta_{34}x^2+\beta_{35}xy+\beta_{36}y^2+\beta_{37}x^3+\beta_{38}x^2y+\beta_{39}xy^2+\beta_{40}y^3) \end{aligned} \tag{2-92a}$$

$$\begin{aligned} \sigma_y=&(\beta_{11}+\beta_{12}x+\beta_{13}y+\beta_{14}x^2+\beta_{15}xy+\beta_{16}y^2+\beta_{17}x^3+\beta_{18}x^2y+\beta_{19}xy^2+\beta_{20}y^3) \\ &+\zeta(\beta_{41}+\beta_{42}x+\beta_{43}y+\beta_{44}x^2+\beta_{45}xy+\beta_{46}y^2+\beta_{47}x^3+\beta_{48}x^2y+\beta_{49}xy^2+\beta_{50}y^3) \end{aligned} \tag{2-92b}$$

$$\begin{aligned} \sigma_{xy}=&(\beta_{21}+\beta_{22}x+\beta_{23}y+\beta_{24}x^2+\beta_{25}xy+\beta_{26}y^2+\beta_{27}x^3+\beta_{28}x^2y+\beta_{29}xy^2+\beta_{30}y^3) \\ &+\zeta(\beta_{51}+\beta_{52}x+\beta_{53}y+\beta_{54}x^2+\beta_{55}xy+\beta_{56}y^2+\beta_{57}x^3+\beta_{58}x^2y+\beta_{59}xy^2+\beta_{60}y^3) \end{aligned} \tag{2-92c}$$

$$\sigma_{xz}=t\left\{(\beta_{61}+\beta_{62}x+\beta_{63}y+\beta_{64}x^2+\beta_{65}xy+\beta_{66}y^2)-\int\left(\frac{\partial\sigma_x}{\partial x}+\frac{\partial\sigma_{xy}}{\partial y}\right)d\zeta\right\} \tag{2-92d}$$

$$\sigma_{yz}=t\left\{(\beta_{67}+\beta_{68}x+\beta_{69}y+\beta_{70}x^2+\beta_{71}xy+\beta_{72}y^2)-\int\left(\frac{\partial\sigma_{xy}}{\partial x}+\frac{\partial\sigma_{y}}{\partial y}\right)d\zeta\right\} \tag{2-92e}$$

$$\sigma_{z}=t\left\{t(\beta_{73}+\beta_{74}x+\beta_{75}y)-\int\left(\frac{\partial\sigma_{xz}}{\partial x}+\frac{\partial\sigma_{yz}}{\partial y}\right)d\zeta\right\} \tag{2-92f}$$

This assumed stress field can be written as

$$\boldsymbol{\sigma}=[P]\ \boldsymbol{\beta} \tag{2-93}$$

This field contains 75 stress parameters β. The number of the stress parameters can be reduced to 67 in order to avoid excessive element stiffening (locking) in the thin plate limit [2.56]. The remaining 67β stress field can be expressed in the form (after a complete renumbering of the stress parameters),

$$\begin{aligned}\sigma_x=&\beta_{16}+[\tfrac{1}{2}(\beta_1-\beta_{53})-\beta_{26}]x+\beta_{17}y+[\tfrac{1}{4}(\beta_2-\beta_{54})-\tfrac{1}{2}\beta_{28}]x^2\\&+[\tfrac{1}{2}(\beta_3-\beta_{55})-2\beta_{29}]xy+\beta_{18}y^2+[\tfrac{1}{6}(\beta_4-\beta_{56})-\tfrac{1}{3}\beta_{31}]x^3\\&+[\tfrac{1}{4}(\beta_5-\beta_{57})-\beta_{32}]x^2y+\beta_{19}y^3+[\tfrac{1}{2}(\beta_6-\beta_{58})-3\beta_{33}]xy^2\\&+\zeta\{\beta_{34}+\beta_{35}x+\beta_{36}y+\beta_{37}x^2+\beta_{38}xy+\beta_{39}y^2\\&+[-\tfrac{3}{4}(\beta_{67}-\beta_{15}+\beta_5+\beta_{57}+2\beta_{12}+2\beta_{64})]x^2y+\beta_{40}xy^2\}\end{aligned} \tag{2-94a}$$

$$\begin{aligned}\sigma_y=&\beta_{20}+\beta_{21}x+[\tfrac{1}{2}(\beta_7-\beta_{59})-\beta_{25}]y+\beta_{22}x^2+[\tfrac{1}{2}(\beta_8-\beta_{60})-2\beta_{27}]xy\\&+[\tfrac{1}{4}(\beta_9-\beta_{61})-\tfrac{1}{2}\beta_{28}]y^2+\beta_{23}x^3+[\tfrac{1}{2}(\beta_{10}-\beta_{62})-3\beta_{30}]x^2y\\&+[\tfrac{1}{4}(\beta_{11}-\beta_{63})-\beta_{31}]xy^2+[\tfrac{1}{6}(\beta_{12}-\beta_{64})-\tfrac{1}{3}\beta_{32}]y^3\\&+\zeta\{\beta_{41}+\beta_{42}x+\beta_{43}y+\beta_{44}x^2+\beta_{45}xy+\beta_{46}y^2+\beta_{47}x^2y\\&+[-\tfrac{3}{4}(\beta_{66}-\beta_{14}+2\beta_4+2\beta_{56}+\beta_{11}+\beta_{63})]xy^2\}\end{aligned} \tag{2-94b}$$

$$\begin{aligned}\sigma_{xy}=&\beta_{24}+\beta_{25}x+\beta_{26}y+\beta_{27}x^2+\beta_{28}xy+\beta_{29}y^2+\beta_{30}x^3+\beta_{31}x^2y\\&+\beta_{32}xy^2+\beta_{33}y^3\\&+\zeta\{\beta_{48}+\beta_{49}x+\beta_{50}y+\beta_{51}x^2+[-\tfrac{3}{4}(\beta_{65}-\beta_{13}+\beta_2+\beta_{54}+\beta_9+\beta_{61})\\&-(\beta_{37}+\beta_{46})]xy+\beta_{52}y^2\}\end{aligned} \tag{2-94c}$$

$$\begin{aligned}\sigma_{yz}=t\{&\frac{1}{2}(1-\zeta)(\beta_7+\beta_8x+\beta_{10}x^2+\beta_{12}y^2)+\frac{1}{8}(1-\zeta)(1-3\zeta)\beta_9y\\&+\frac{1}{4}(1-\zeta)(-1-3\zeta)\beta_{11}xy+\frac{1}{2}(1+\zeta)(\beta_{59}+\beta_{60}x+\beta_{62}x^2+\beta_{64}y^2)\\&+\frac{1}{8}(1+\zeta)(1+3\zeta)\beta_{61}y+\frac{1}{4}(1+\zeta)(-1+3\zeta)\beta_{63}xy+\frac{1}{2}(1-\zeta^2)\{(\beta_{43}\\&+\beta_{49})+(\beta_{45}+2\beta_{51})x+[-\frac{3}{4}(\beta_{65}-\beta_{13}+\beta_2+\beta_{54})+(\beta_{46}-\beta_{37})]y\\&\quad+\beta_{47}x^2+[-\frac{3}{2}(\beta_{66}-\beta_{14})-3(\beta_4+\beta_{56})]xy\}\}\end{aligned} \tag{2-94d}$$

$$\begin{aligned}\sigma_{zx}=t\{&\frac{1}{2}(1-\zeta)(\beta_1+\beta_3y+\beta_4x^2+\beta_6y^2)+\frac{1}{8}(1-\zeta)(1-3\zeta)\beta_2x\\&+\frac{1}{4}(1-\zeta)(-1-3\zeta)\beta_5xy+\frac{1}{2}(1+\zeta)(\beta_{53}+\beta_{55}y+\beta_{56}x^2+\beta_{58}y^2)\\&\quad+\frac{1}{8}(1+\zeta)(1+3\zeta)\beta_{54}x+\frac{1}{4}(1+\zeta)(-1+3\zeta)\beta_{57}xy\\&+\frac{1}{2}(1-\zeta^2)\{(\beta_{35}+\beta_{50})+[-\frac{3}{4}(\beta_{65}-\beta_{13}+\beta_9+\beta_{61})+(\beta_{37}-\beta_{46})]x\\&+(\beta_{38}+2\beta_{52})y+[-\frac{3}{2}(\beta_{67}-\beta_{15})-3(\beta_{12}+\beta_{64})]xy+\beta_{40}y^2\}\}\end{aligned} \tag{2-94e}$$

$$\begin{aligned}\sigma_z=t^2\{&\frac{1}{4}(1-\zeta)^2(2+\zeta)(\beta_{13}+\beta_{14}x+\beta_{15}y)+\frac{1}{4}(1+\zeta)^2(2-\zeta)(\beta_{65}+\beta_{66}x+\beta_{67}y)\\&-\frac{1}{4}(1-\zeta)^2(1+\zeta)[(\beta_2+\beta_9)+(2\beta_4+\beta_{11})x+(\beta_5+2\beta_{12})y]\\&+\frac{1}{4}(1-\zeta)(1+\zeta)^2[(\beta_{54}+\beta_{61})+(2\beta_{56}+\beta_{63})x+(\beta_{57}-2\beta_{64})y]\}\end{aligned} \tag{2-94f}$$

This assumed stress field satisfies the equilibrium equations, but the number of the stress parameters in the field is very large. Furthermore, the resulting element may be not free from kinematic deformation modes.

c) Stress Modes Matched with Strain Modes

In the equilibrating stress field above, the complete linear or cubic polynomials are used. It usually causes the overabundance of stress parameters. On the other hand, reducing the number of stress parameters may cause spurious kinematic deformation modes in the hybrid element. One procedure to suppress any spurious kinematic deformation modes is to match one stress mode individually with one strain mode which represents a basic deformation mode such that the strain energy caused by each strain mode is not zero [2.47].

Within an element with n degrees of freedom and r rigid body modes, the displacement distribution in the element can be represented by n-r basic deformation modes and r rigid body modes. If the basic deformation modes of the element are expressed in the form

$$\mathbf{u}_i = [\overline{N_i}]\, \boldsymbol{\alpha}_i \qquad i=1,2,\ldots n-r \tag{2-95}$$

the individual basic strain modes can be expressed as

$$\overline{\boldsymbol{\varepsilon}_i} = [D]\, [\overline{N_i}]\, \boldsymbol{\alpha}_i \qquad i=1,2,\ldots n-r \tag{2-96}$$

Moreover, if the individual stress mode is expressed in the form

$$\overline{\boldsymbol{\sigma}_i} = [\sigma_i]\, \boldsymbol{\beta}_i \tag{2-97}$$

Then, it is assumed that, for each deformation mode, there is one stress mode such that the strain energy

$$I_i = \boldsymbol{\beta}_i^T \{ \int_V [\sigma_i]^T ([D]\, [\overline{N_i}])\, dV \}\, \boldsymbol{\alpha}_i \tag{2-98}$$

is non-zero.

1. 2-D, 4-node Plane Element

Within a 4-node plane element, the assumed displacement field (2-82) is

$$\begin{aligned} u &= a_0 + a_1\xi + a_2\eta + a_3\xi\eta \\ v &= b_0 + b_1\xi + b_2\eta + b_3\xi\eta \end{aligned} \tag{2-99}$$

There are eight displacement degrees of freedom (n=8) and three rigid body modes (r=3). Therefore, there are five basic deformation modes (n-r=5). Corresponding to the basic deformation modes, the basic strains are

$$\begin{aligned} \varepsilon_x &= \alpha_1 + \alpha_4\eta \\ \varepsilon_y &= \alpha_2 + \alpha_5\xi \\ \varepsilon_{xy} &= 2\alpha_3 \end{aligned} \tag{2-100}$$

In which, each α_i is associated with a distinct strain mode. It can be written as follows,

$$\boldsymbol{\varepsilon}=\alpha_1\begin{Bmatrix}1\\0\\0\end{Bmatrix}+\alpha_2\begin{Bmatrix}0\\1\\0\end{Bmatrix}+\alpha_3\begin{Bmatrix}0\\0\\2\end{Bmatrix}+\alpha_4\begin{Bmatrix}\eta\\0\\0\end{Bmatrix}+\alpha_5\begin{Bmatrix}0\\\xi\\0\end{Bmatrix} \tag{2-101}$$

The deformation energy due to assumed stress modes and basic deformation modes is calculated by

$$I_i=\int_{-1}^{1}\int_{-1}^{1}\boldsymbol{\sigma}_i^T\boldsymbol{\varepsilon}_i\,d\xi\,d\eta \tag{2-102}$$

Using the energy constraint, $I_i \neq 0$, the corresponding stress modes can be found,

$$\boldsymbol{\sigma}=\beta_1\begin{Bmatrix}1\\0\\0\end{Bmatrix}+\beta_2\begin{Bmatrix}0\\1\\0\end{Bmatrix}+\beta_3\begin{Bmatrix}0\\0\\2\end{Bmatrix}+\beta_4\begin{Bmatrix}\eta\\0\\0\end{Bmatrix}+\beta_5\begin{Bmatrix}0\\\xi\\0\end{Bmatrix} \tag{2-103}$$

This assumed stress field can be written in the form,

$$\begin{Bmatrix}\sigma_x\\\sigma_y\\\sigma_{xy}\end{Bmatrix}=\begin{bmatrix}1&0&0&\eta&0\\0&1&0&0&\xi\\0&0&1&0&0\end{bmatrix}\begin{Bmatrix}\beta_1\\\beta_2\\\cdot\\\cdot\\\beta_5\end{Bmatrix} \tag{2-104}$$

2. 3-D, 8-node Brick Element

This method can be also used to determine stress modes for three dimensional 8-node brick element. Within the element, there are twenty-four displacement degrees of freedom (n=24) and six rigid body modes (r=6). Therefore, there are eighteen basic deformation modes (n-r=18). They are [2.47]

$$
\begin{aligned}
u&=\alpha_1\xi+\frac{1}{2}\alpha_4\eta+\frac{1}{2}\alpha_6\zeta+\alpha_7\xi\eta+\alpha_8\zeta\xi+\alpha_9\xi\eta\zeta+\alpha_{16}\eta\zeta \\
v&=\alpha_2\eta+\frac{1}{2}\alpha_4\xi+\frac{1}{2}\alpha_5\zeta+\alpha_{10}\xi\eta+\alpha_{11}\eta\zeta+\alpha_{12}\xi\eta\zeta+\alpha_{17}\zeta\xi \\
w&=\alpha_3\zeta+\frac{1}{2}\alpha_5\eta+\frac{1}{2}\alpha_6\xi+\alpha_{13}\eta\zeta+\alpha_{14}\zeta\xi+\alpha_{15}\xi\eta\zeta+\alpha_{18}\xi\eta
\end{aligned}
\tag{2-105}
$$

in which, each α_i is associated with a basic deformation mode. The deformation energy due to assumed stresses modes and basic deformation modes is calculated by

$$
I_i=\int_{-1}^{1}\int_{-1}^{1}\int_{-1}^{1}\boldsymbol{\sigma}_i^T[DN_i]\,\alpha_i\,d\xi\,d\eta\,d\zeta \tag{2-106}
$$

Using the energy constraint again, $I_i \neq 0$, the stress modes corresponding to the basic deformation modes can be found and the assumed stress field can be expressed in the form,

$$
\boldsymbol{\sigma}=\begin{bmatrix}
1&0&0&0&0&0&\eta&\zeta&\eta\zeta&0&0&0&0&0&0&0&0&0\\
0&1&0&0&0&0&0&0&0&\xi&\zeta&\zeta\xi&0&0&0&0&0&0\\
0&0&1&0&0&0&0&0&0&0&0&0&\eta&\xi&\xi\eta&0&0&0\\
0&0&0&1&0&0&0&0&0&0&0&0&0&0&0&0&\zeta&0\\
0&0&0&0&1&0&0&0&0&0&0&0&0&0&0&0&0&\xi\\
0&0&0&0&0&1&0&0&0&0&0&0&0&0&0&\eta&0&0
\end{bmatrix}
\begin{Bmatrix}\beta_1\\ \beta_2\\ \cdot\;\cdot\\ \beta_{18}\end{Bmatrix}
\tag{2-107}
$$

Using this method, the assumed stress field contains minimum number of stress modes (or stress parameters), and the resulting element is free from any kinematic deformation modes. But, in this procedure, the basic deformation modes of the element must be first found, and then the stress modes can be obtained by checking if deformation energy is equal to zero. For many elements, it is difficult to find the basic deformation modes at first. Therefore, the use of the method is limited.

2.5 DIFFICULTIES WITH THE HYBRID FINITE ELEMENT METHOD

In general, the hybrid stress element has two important disadvantages: the presence of spurious kinematic deformation modes and the inversion of the flexibility matrix [H].

Because the assumed stress field of the conventional hybrid elements contains six stress components, there always exist a large number of stress parameters in the stress field. So, the inversion of the flexibility matrix is the most costly operation. For instance, an assumed stress field may contain more than fifty stress parameters for the stress analysis of 3-D structures and hundreds of stress parameters for the analysis of composite structure [2.56]. It suggests a poor performance in terms of computing time when compared with the single-field displacement models. However, this limitation can be overcome by reducing the number of stress components in the assumed stress field of conventional hybrid elements for the analysis of composite structures.

For analysis of composite structures, it is not necessary to introduce all components of stresses into an assumed stress field. Although all components of displacement, strain and stress must be continuous within each layer of a laminated composite, only the in-plane derivatives ε_x , ε_y , γ_{xy} and transverse stresses σ_z , τ_{xz} ,τ_{yz} must be continuous at the layer interface with prefect bonding. Therefore, the main requirement in developing finite element is to satisfy all of the continuity conditions on displacements and transverse stresses at interlaminar surfaces and traction-free condition on the upper and lower surfaces. In order to enforce the transverse stress continuity, it is needed only to introduce three transverse stresses into the assumed stress field [2.58]. This motivates researchers to develop new variational principle for new types of hybrid elements.

In view of the efficiency of finite element model, on the one hand, the number of the stress parameters (or stress modes) in assumed stress field should be reduced to as small as possible. According to the necessary condition (2-78), the minimum number of stress parameters (or stress modes) may equal m (= n - r). On the other hand, there are many examples indicating that there are spurious kinematic deformation modes in the hybrid elements when the requirement (2-78) is satisfied. In order to suppress these kinematic deformation modes, it is proposed to add stress modes of high order terms in thc assumed stress field. This means to increase the number of stress parameters in the stress field. Therefore, the question is that how many and what kind of stress modes must be introduced into the assumed stress field. An ideal situation is that an assumed stress field contains m (=n-r) least-order stress modes and its resulting element is free from kinematic deformation modes. This kind of assumed stress fields is considered to be best and is optimal with respect to computer resources. The procedure to derive this optimal stress field is presented in Chapter 3.

In this chapter, it is also shown that the assumed stress field of a hybrid stress element can be constructed by various approaches. Thus, a hybrid stress element may have many different assumed stress fields. However the relationship among them has not been revealed due to the fact that the nature of the assumed stress field has not been investigated sufficiently. In order to develop a rational way for deriving the optimal assumed stress field, it will be necessary further to study the stress modes in the assumed stress field. This will be presented also in Chapter 3.

REFERENCES

2.1 M. Turner, R. Clough, R. Matrin, and L. Topp, 'Stiffness and deflection analysis of complex structures', J, Aeronaut Sci., vol. 23, no. 9, 805-823(1956).

2.2 R. Clough, 'The finite element method in plane stress analysis', Proc. 2nd ASCE Conf. Electronic Comp., Pittsburg, 1960.

2.3 A. Adini and R. Clough, 'Analysis plate bending by the finite element method', Report to National Science Foundation, 1961.

2.4 A. Adini, 'Analysis of shell structures by the finite element method', Ph. D. thesis, Department of Civil Engineering, University of California, Berkeley, 1961.

2.5 R. Melosh, 'Basis for derivation of matrices by the direct stiffness method', AIAA J., vol. 1, (1963).

2.6 J. Argyris, 'Matrix analysis of three-dimensional elastic media -- small and large displacements', AIAA J., vol. 3, 45-51(1965).

2.7 J. Tocher and B. Hartz, 'High-order finite element for plane stress', ASCE 93, no. EM4, 1967.

2.8 T.H.H. Pian, 'Derivation of element stiffness matrices by assumed stress distributions', AIAA J. vol. 2, 1333-1336(1964).

2.9 P. Tong and T.H.H. Pian, 'A variational principle and the convergence of a finite element method based on assumed stress distribution', Int. J. Solids Struct., vol. 5, 463-472(1969).

2.10 S. N. Atluri, R. H. Gallagher, and O. C. Zienkiewicz, eds., Hybrid and Mixed Finite Element Models, John Wiley, 1983.

2.11 G. F. Carey and J. T. Oden, Finite Elements: A Second Course, vol. II, Prentice-Hall, 1982.

2.12 T. H. H. Pian and P. Tong, 'Basis of finite element methods for solid continua', Int. J. Numer. Methods Engrg., vol. 1, 3-28(1969).

2.13 A. Poceski, Mixed Finite Element Method, Springer-Verlag Berlin Heidelberg, 1992.

2.14 F. Brezzi and M. Fortin, Mixed and Hybrid Finite Element Methods, Springer-Verlag New York Inc. 1991.

2.15 T. H. H. Pian, 'A historical note about hybrid elements', Int. J. Numer. Methods Engrg., vol. 12, 891-892(1978).

2.16 T. H. H. Pian, 'Recent advances in hybrid/mixed finite elements', in:G.Q. He and Y. K. Cheung (eds.), Proc. Conf. on Finite Element Methods, Shanghai, China, 82-89(1982).

2.17 T. H. H. Pian, 'Thirty-year history of hybrid stress finite element methods', in: P. K. K. Lee and L. G. Tham (eds.), Y. K. Cheung Symposium, 15 December 1994, University of Hong Kong, 23-31(1994).

2.18 T. H. H. Pian, 'State-of-the-art development of hybrid/mixed finite element method', Finite Elements in Analysis and Design, vol. 21 5-20(1995).

2.19 S. T. Mau, P. Tong, and T. H. H. Pian, 'Finite element solutions for laminated thick plates', J. Composite Materials, vol. 6, 304-311(1972).

2.20 R. L. Spilker, S. C. Chou, and O. Orringer, 'Alternate hybrid-stress elements for analysis of multilayer composite plates', J. Composite Materials, vol. 11, 51-70(1977).

2.21 R. L. Spilker, 'A hybrid stress finite element formulation for thick multilayer laminates', Computers & Structures, vol. 11, 507-514(1980).

2.22 W.-J. Liou & C. T. Sun, 'A three dimensional hybrid stress isoparametric element for the analysis of laminated composited plates', Computers & Structures, vol. 25, no. 2, 241-249(1987).

2.23 K. Washizu, Variational Methods in Elasticity and Plasticity, Pergamon Press, London, 1968.

2.24 W. Z. Chien, Generalized Variational Principles, Knowledge Press, Shanghai, 1985.

2.25 H. C. Hu, 'On some variational principles in the theory of elasticity and plasticity', Scientia Sinica,

Beijing, vol. 4, no. 1, 33-54(1955).

2.26 K. Washizu, On the Variational Principles of Elasticity and Plasticity, Aeroelastic and Structure Research Laboratory, Massachusetts Institute of Technology, Technical Report 25-18, March 1955.

2.27 E. Hellinger, Der allgemeine Ansatz der Mechanik der Kontinua, Encyclopadie der Mathematischen Wissenschaften, vol. 4, Part 4, 602-694(1914).

2.28 E. Reissner, 'On a variational theorem in elasticity', J. of Mathematics and Physics, vol. 29, no. 2, 90-95(1950).

2.29 W. Z. Chien, 'Method of high-order Lagrange multipliers and generalized variational principles of elasticity with more general forms of functionals', Applied Mathematics and Mechanics, vol. 4, 143-157(1983).

2.30 C. A. Fellppa and C. Militello, 'Variational formulation of high-performance finite elements: parametrized variational principles', Computers & Structures, vol. 36, no. 1, 1-11(1990).

2.31 C. A. Felippa, 'Parametric unification of matrix structural analysis: classical formulation and d-connected mixed elements', Finite Elements in Analysis and Design, vol. 21, 45-74(1995).

2.32 Q. Huang, The Formulation of Hybrid Displacement Finite Elements, Shanghai Institute of Applied Mathematics and Mechanics, Shanghai University, Technical Report, 1990.

2.33 T. T. Pian, 'Variational principles for finite element methods in solid mechanics', in: Z. M. Zheng (ed.), Applied Mechanics, International Academic Publishers, Beijing, 1989.

2.34 T. Belytschko and J. S. S. Ong, 'A consistent control of spurious single modes in the 9-node Lagrange element for the Laplace and Mindlin Plate Equations', Comput. Methods Appl. Mech. Eng., vol. 44, 269-295(1984).

2.35 R. D. Cook, 'Modified formulations for nine-d.o.f. plane triangles that include vertex rotation', Int. J. Numer. Methods Eng., vol. 31, 825-835(1991).

2.36 C. A. Felippa, 'Parametrized multifield variational principles in elasticity: I. mixed functionals, II. hybrid functional and the free formulation', Commun. Appl. Numer. Methods, vol. 5, 79-98(1989).

2.37 K. T. Kavanagh and S. W. Key, 'A note on selective and reduced integration techniques in finite element methos', Int. J. Numer. Methods Eng., vol. 4, 148-150(1972).

2.38 K. Y. Sze, 'Finite element formulation by parametrized hybrid variational principles: variable stiffness and removal of locking', Int. J. Numer. Methods Eng., vol. 37, 2797-2818(1994).

2.39 R. D. Henshell, 'On hybrid finite elements', Proceeding of the Brunel University Conference of the Institute of Math and Its Application, April, 1972.

2.40 B. M. Fraeijs de Veubake, Proceeding of the Conference on Matrix Methods in Structure Machanics, AFFDL-TR-66-80, (1966).

2.41 F. Brezzi, 'On the existence, uniqueness and approximation of saddle point problems arising from Lagrange multiplers', RAIRO 8(R2), 129-151(1974).

2.42 I. Babuska, J. T. Oden, and J. K. Lee, 'Mixed-hybrid finite element approximations of second-order elliptic boundary-value problems, Part I ', Computer Methods in Applied Mechanics and Engineering, vol. 11, 175-206(1977).

2.43 B. Fraeijs de Veubeke, 'Displacement and equilibrium models in the finite element methods', in: O. C. Zienkiewicz and G. S. Holister, (eds.), Stress Analysis, John Wiley, London, 1965.

2.44 H. Stolarski and T. Belytschko, 'Limitation principles for mixed finite elements based on the Hu-Washizu variational formulation', Comput. Meth. Appl. Mech. Engng. vol. 60, 195-216(1987).

2.45 E. F. Punch and S. N. Atluri, 'Development and testing of stable, invariant, isoparametric curvilinear 2- and 3-D hybrid-stress elements', Computer Methods in Applied Mechanics and Engineering, vol. 47, 331-356(1984).

2.46 R. Rubinstein, E. F. Punch, and S. N. Atluri, 'An analysis of, and remedies for, kinematic modes in hybrid-stress finite elements: selection of stable, invariant stress fields', Computer Methods in Applied Mechanics and Engineering, vol. 38, 63-92(1983).

2.47 T. H. H. Pian and D. P. Chen, 'On the suppression of zero energy deformation modes', International Journal for Numerical Methods in Engineering, vol. 19, 1741-1752(1983).

2.48 T. H. H. Pian and P. Tong, 'Relations between incompatible displacement model and hybrid stress model', International Journal for Numerical Methods in Engineering, vol. 22, 173-181(1986).

2.49 T. H. H. Pian and C. C. Wu, 'A rational approach for choosing stress terms for hybrid finite element formulations', International Journal for Numerical Methods in Engineering, vol. 26, 2331-

2343(1988).
2.50 C. C. Wu and Y. K. Cheung, 'On optimization approaches of hybrid stress elements', Finite Elements in Analysis and Design, vol. 21, 111-128(1995).
2.51 K. Y. Sze, C. L. Chow, and W. J. Chen, 'A rational formulation of iso-parametric hybrid stress element for three dimensional stress analysis', Finite Elements in Analysis and Design, vol. 7, 61-72(1990).
2.52 W. J. Chen and Y. K. Cheung, 'Three-dimensional 8-node and 20-node refined hybrid isoparametric elements', International Journal for Numerical Methods in Engineering, vol. 35, 1871-1889(1992).
2.53 K. Y. Sze, 'Control of spurious mechanisms for 20-node and transition sub-integrated hexahedral elements', International Journal for Numerical Methods in Engineering, vol. 37, 2235-2250(1994).
2.54 K. Y. Sze, 'A novel approach for devising higher-order hybrid elements', International Journal for Numerical Method in Engineering, vol. 36, 3303-3316(1993).
2.55 K. Y. Sze and C. L. Chow, 'Efficient hybrid/mixed elements using admissible matrix formulation', Conputer Methods in Applied Mechanics and Engineering, vol. 99, 1-26(1992).
2.56 R. L. Spilker, 'Hybrid-stress eight-node elements for thin and thick multilayer laminated plates', International Journal for Numerical Methods in Engineering, vol. 18, 801-828(1982).
2.57 R. L. Spilker and N. I. Munir, 'The hybrid-stress model for thin plates', International Journal for Numerical Methods in Engineering, vol. 15, 1239-1260(1980).
2.58 S. V. Hoa and W. Feng, 'Finite elements for analysis of composite structures', in: S. V. Hoa (ed.), Computer-Aided Desing of Polymer-Matrix Composite Structures, 289-359, Marcel Dekker, Inc., 1995.

Chapter 3

DEVELOPMENT OF HYBRID ELEMENT TECHNIQUE FOR ANALYSIS OF COMPOSITES

3.1 INTRODUCTION

The previous two chapters presented a brief overview of the displacement finite element formulation and the hybrid finite element formulation. These finite element formulations have been used for the analysis of structures made of isotropic materials. They also have been used for the analysis of laminated composite structures. However, the results for laminated composite structures need improvement due to the fact that there are many levels of discontinuities in the laminated composites. These discontinuities give rise to many regions of high stress gradients. On the microstructural level, there is discontinuity in material properties as one moves from fiber to matrix or vice versa. For the purpose of calculation at the lamina level, the fiber and matrix properties are averaged over an effective unit cell and the effective modulus approach is used for macromechanics. The average properties of individual lamina are usually obtained based on this assumption. Moreover, when many laminae are stacked to form laminates, due to the variation in fiber orientation from lamina to lamina, the interlaminar stresses occur near the interfaces between the laminae. The interlaminar failure modes caused by the interlaminar stresses are major failure modes in laminates because interlaminar strengths are usually orders of magnitude smaller than intralaminar strengths. This problem has been with designers and researchers for the past thirty years. Many numerical techniques have been proposed, the majority of them using the finite element method [3.1]. However until the present time, the problem has not been resolved satisfactorily. The main difficulty is in the efficiency in obtaining transverse stresses accurately. Without efficient means to obtain accurate transverse stresses, it is difficult to obtain efficient ways to predict interlaminar failure.

In finite element method, the conventional displacement finite elements work well with the stress analysis of homogenous materials. However, they have an inherent disadvantage that differentiation has to be performed on the approximated displacements to obtain strains and subsequently stresses. The accuracy will deteriorate because the numerical differentiation of approximate quantities tend to magnify the errors. In addition, the convergence of displacement finite element model for problems with high stress gradients is slow. For composite laminates, the interfaces between layers are usually locations of high gradients of stress due to discontinuity in material properties as one moves from one layer to the next. Therefore, the use of displacement finite element model requires fine element mesh and extensive amount of computer space and time to be able to determine stresses with any degree of accuracy. This excessive requirement of computer resources has been a deterrent to accurate and efficient stress calculation in composite laminates. Furthermore, the displacement elements can not satisfy well the requirements in analysis of composites. The main requirement in developing finite element for the analysis of composite is to satisfy all of the continuity conditions of displacements and transverse stresses at interlaminar surfaces, and traction-free condition on the upper and/or lower surfaces. As an alternate, the conventional hybrid stress elements [3.2-3.3] have been used to analyze composite structures.

The conventional hybrid elements have the ability to satisfy these conditions exactly. The hybrid element formulation assumes the stress field directly from the beginning. Therefore in hybrid finite element formulation, no differentiation on the approximated values has to be carried out and the degree of accuracy of the stresses is the same as that of the displacements. This is the inherent advantage of the hybrid finite element method. However, the conventional hybrid elements contain six stress components. This will require much computer CPU time due to the presence of many stress parameters β in assumed stress fields and the inversion of the flexibility matrix [H]. In fact, it is not necessary to introduce all components of stresses into an assumed stress field for analysis of composite structures. In order to satisfy the continuity condition of stresses at interlaminar surfaces, three transverse stresses are only needed in the assumed stress field. Therefore, new hybrid finite element techniques have been developed for the stress analysis of composites [3.4-3.15].

3.2 COMPOSITE VARIATIONAL PRINCIPLE

In the development of the partial hybrid finite elements for analysis of composites, the first thing is the identification of globally continuous variables and locally continuous variables. Usually, the lamina plane is denoted by the Cartesian co-ordinates x, y, and the through thickness direction by z (shown in figure 13). In the composite laminates, all components of displacement, strain, and stress are continuous within each layer. At the layer interface with perfect bonding, the displacements are also continuous due to the compatibility condition. As a result, the in-plane derivatives (three inplane strains) ε_x, ε_y, ε_{xy} are continuous across the thickness. On the other hand, the reaction forces give rise to transverse stresses (interlaminar stresses) σ_z, σ_{zx}, σ_{zy}', and they are also continuous across the thickness because of the equilibrium condition. This means, along the thickness of composites, the in-plane strains (ε_x, ε_y, ε_{xy}) and transverse stresses (σ_z, σ_{zx}, σ_{zy}) are globally continuous variables. The other components of strain and stress (transverse strains ε_z, ε_{zx}, ε_{zy} and in-plane stresses σ_x, σ_y, σ_{xy}) are not necessarily continuous across the interfaces between different layers. They are continuous within each layer and are locally continuous variables. Therefore, the *globally continuous variables* are those that are continuous not only within the plane of the laminate but they are also continuous across the interface from one layer to the next. This is the result of consideration for compatibility and equilibrium. The *locally continuous variables* are those that are continuous only within the plane of the lamina but are not necessarily continuous across the interface from one lamina to the next.

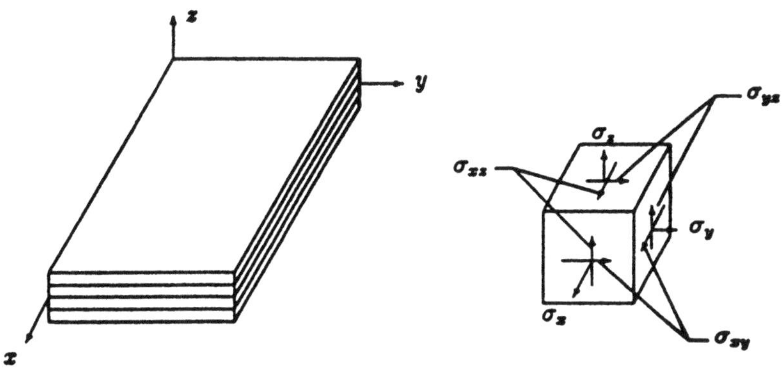

Figure 13 Composite structure and its co-ordinate system

By classifying the variables into these two groups, the stress σ and strain $\boldsymbol{\varepsilon}$ can be divided into the in-plane and transverse parts,

$$\sigma_L = [\ \sigma_x\ \sigma_y\ \tau_{xy}\ ]^T \quad \text{and} \quad \sigma_g = [\ \sigma_z\ \tau_{zy}\ \tau_{zx}\ ]^T$$

$$\boldsymbol{\varepsilon}_g = [\ \boldsymbol{\varepsilon}_x\ \boldsymbol{\varepsilon}_y\ \gamma_{xy}\ ]^T \quad \text{and} \quad \boldsymbol{\varepsilon}_L = [\ \boldsymbol{\varepsilon}_z\ \gamma_{zy}\ \gamma_{zx}\ ]^T \qquad (3-1)$$

Combining the globally continuous variables in composites, the globally continuous vector is defined as

$$\mathbf{q}=\begin{Bmatrix}\boldsymbol{\varepsilon}_g\\ \boldsymbol{\sigma}_g\end{Bmatrix} \tag{3-2}$$

Similarly, combining the locally continuous variables, the locally continuous vector is defined as

$$\mathbf{p}=\begin{Bmatrix}\boldsymbol{\sigma}_L\\ -\boldsymbol{\varepsilon}_L\end{Bmatrix} \tag{3-3}$$

in which the negative sign is introduced to ensure the symmetry of the combined constitutive relation which is

$$\mathbf{p} = [\,R\,]\,\mathbf{q} \tag{3 - 4}$$

or

$$\begin{Bmatrix}\boldsymbol{\sigma}_L\\ -\boldsymbol{\varepsilon}_L\end{Bmatrix}=\begin{bmatrix}R_1 & R_2\\ R_2^T & R_3\end{bmatrix}\begin{Bmatrix}\boldsymbol{\varepsilon}_g\\ \boldsymbol{\sigma}_g\end{Bmatrix} \tag{3-4$'$}$$

where [R] is called the combined constitutive matrix. Because the constitutive relation can be expressed in the form,

$$\boldsymbol{\sigma}=[C]\,\boldsymbol{\varepsilon} \quad or \quad \begin{Bmatrix}\boldsymbol{\sigma}_L\\ \boldsymbol{\sigma}_g\end{Bmatrix}=\begin{bmatrix}C_1 & C_2\\ C_2^T & C_3\end{bmatrix}\begin{Bmatrix}\boldsymbol{\varepsilon}_g\\ \boldsymbol{\varepsilon}_L\end{Bmatrix} \tag{3-5}$$

where [C] is the stiffness matrix of materials, and

$$\boldsymbol{\varepsilon}=[S]\,\boldsymbol{\sigma} \quad or \quad \begin{Bmatrix}\boldsymbol{\varepsilon}_g\\ \boldsymbol{\varepsilon}_L\end{Bmatrix}=\begin{bmatrix}S_1 & S_2\\ S_2^T & S_3\end{bmatrix}\begin{Bmatrix}\boldsymbol{\sigma}_L\\ \boldsymbol{\sigma}_g\end{Bmatrix} \tag{3-5$'$}$$

where [S] is the compliance matrix of materials, the matrix [R] can be expressed as

$$[R]=\begin{bmatrix} R_1 & R_2 \\ R_2^T & R_3 \end{bmatrix}=\begin{bmatrix} C_1-C_2C_3^{-1}C_2^T & C_2C_3^{-1} \\ C_3^{-1}C_2^T & -C_3^{-1} \end{bmatrix} \tag{3-6}$$

and

$$[R]=\begin{bmatrix} R_1 & R_2 \\ R_2^T & R_3 \end{bmatrix}=\begin{bmatrix} S_1^{-1} & -S_1^{-1}S_2 \\ -S_2^TS_1^{-1} & S_2^TS_1^{-1}S_2-S_3 \end{bmatrix} \tag{3-6)'$$

Due to the fact that the matrices [C] and [S] are symmetric matrices, the matrix [R] can be proven to be a symmetric matrix. It can be shown that

$$[R]^T=[R] \tag{3-7}$$

Thus, an elasticity problem for composite structures can be described.

Elasticity Problem for Composite Structures

Consider a linear anisotropic elastic body under static loading. The body occupies the volume V and is bounded by the surface S, which is decomposed into S: $S_d \cup S_t$. Boundary displacements are prescribed on S_d, whereas surface tractions are prescribed on S_t. The outward unit normal on S is denoted by **n**. The following relations between three fields: globally continuous vector **q**, locally continuous vector **p**, and displacement **u** in the volume have to be satisfied.

1. the partial strain-displacement equations:

$$\varepsilon_{ij}=\frac{1}{2}(u_{i,j}+u_{j,i}) \qquad i,j=1,2 \tag{3-8}$$

or

$$\boldsymbol{\varepsilon_g}=\boldsymbol{D_g u} \tag{3-8)'$$

2. the partial strain-displacement relations:

$$\varepsilon_{ij}=\frac{1}{2}(u_{i,j}+u_{j,i}) \qquad i=1,2,3 \quad j=3 \tag{3-9}$$

or

$$\boldsymbol{\varepsilon}_L=\boldsymbol{D}_L\boldsymbol{u} \tag{3-9}$$

3. the stress-strain equations (constitutive equations):

$$p = [R] q \tag{3-10}$$

4. the equilibrium equations:

$$\sigma_{ij,j}+F_i=0 \tag{3-11}$$

in which, **F** is the body force in V.

Moreover, there are three sets of boundary conditions for the displacement field and stress field.

5. the traction boundary conditions:

$$\boldsymbol{\sigma}\cdot\boldsymbol{n}=\boldsymbol{T}_n \quad and \quad \boldsymbol{T}_n=\boldsymbol{T} \tag{3-12}$$

or

$$\sigma_{ij}n_j=T_{ni} \quad and \quad T_{ni}=T_i \qquad on\ S_t \tag{3-12'}$$

in which, **T** is the prescribed surface force on S_t.

6. the displacement boundary conditions:

$$\boldsymbol{u}=\boldsymbol{d} \tag{3-13}$$

or

$$u_i = d_i \quad \text{on} \quad S_d \tag{3-13)'$$

in which, **d** is the prescribed displacement on S_d.

7. the conjunction conditions at interlayer surfaces:

$$\begin{array}{ll} u_i^k = u_i^{k+1} & i=1,2,3 \\ \sigma_{i3}^k = \sigma_{i3}^{k+1} & k=1,2,...,N \end{array} \tag{3-14}$$

where N is the number of layers along the thickness of composite structures.

Composite Energy

For an elastic body, the potential energy can be expressed as a quadratic form of strains,

$$A(\boldsymbol{u}) = \frac{1}{2} \boldsymbol{\varepsilon}^T [C] \boldsymbol{\varepsilon} \tag{3-15}$$

and the complementary energy can be expressed as a quadratic form of stresses,

$$B(\boldsymbol{u}) = \frac{1}{2} \boldsymbol{\sigma}^T [S] \boldsymbol{\sigma} \tag{3-16}$$

Similarly, one can define a new energy, named *composite energy,* as a quadratic form of the globally continuous variables due to the fact that the constitutive matrix [R] is symmetric matrix (3-7),

$$E(\boldsymbol{q}) = \frac{1}{2} \boldsymbol{q}^T [R] \boldsymbol{q} \tag{3-17}$$

or

$$E=\frac{1}{2}\sum_{i=1}^{6}\sum_{j=1}^{6}R_{ij}q_iq_j \tag{3-18}$$

Thus, the constitutive equations (3-10) can be written in the form,

$$\frac{\partial E(\boldsymbol{q})}{\partial q_i}=p_i \tag{3-10}$$

Composite Energy Functional

The variational functional based on the composite energy can be derived through different ways. Huang [3.5] established the variational functional by weighted residual method, Reissner [3.17-3.18] developed it using Lagrange multiplier and 'partial' Legendre transformation method, and Moriya [3.19] developed it through the Hu-Washizu variational principle. Lately, Pian [3.13] used the Hellinger-Reissner variational principle to obtain the functional. In view of simplicity, the variational functional, termed *composite energy functional,* is presented by means of the Hellinger-Reissner variational principle, and their difference is revealed.

The Hellinger-Reissner variational principle (see chapter 2) contains two fields: displacement field and stress field. The constraint conditions are constitutive equations (2-2) and displacement boundary conditions (2-5) only. The strain-displacement equations (2-1), equilibrium equations (2-3), and traction boundary conditions (2-4) are only satisfied a posteriori. Satisfying the displacement boundary conditions (2-5) a priori, the variational functional (2-25) can be expressed as follows,

$$\Pi_{III}=\int_V\left[-\frac{1}{2}\boldsymbol{\sigma}^T[S]\,\boldsymbol{\sigma}+\boldsymbol{\sigma}^T(\boldsymbol{D}\mathbf{u})-\boldsymbol{F}^T\mathbf{u}\right]dV-\int_{S_t}\boldsymbol{T}^T\mathbf{u}\,dS \tag{2-25}$$

By means of the definition (3-1), (3-8), and (3-9), one has

$$\boldsymbol{\sigma}=\begin{Bmatrix}\boldsymbol{\sigma}_L\\ \boldsymbol{\sigma}_g\end{Bmatrix} \quad and \quad \boldsymbol{D}\mathbf{u}=\begin{Bmatrix}\boldsymbol{D}_g\mathbf{u}\\ \boldsymbol{D}_L\mathbf{u}\end{Bmatrix} \tag{3-19}$$

Substituting them into the (2-25), the functional is written in the form,

$$\Pi_{III}=\int_V [-\frac{1}{2}[\boldsymbol{\sigma}_L^T\ \boldsymbol{\sigma}_g^T]\ [S]\begin{Bmatrix}\boldsymbol{\sigma}_L\\ \boldsymbol{\sigma}_g\end{Bmatrix}+[\boldsymbol{\sigma}_L^T\ \boldsymbol{\sigma}_g^T]\begin{Bmatrix}\boldsymbol{D}_g\mathbf{u}\\ \boldsymbol{D}_L\mathbf{u}\end{Bmatrix} -\boldsymbol{F}^T\mathbf{u}]\ dV-\int_{S_t}\boldsymbol{T}^T\mathbf{u}\ dS \tag{3-20}$$

Using the constitutive equations (3-6)' and (3-4)' to eliminate the in-plane stress σ_L, the first term in the functional becomes

$$\int_V -\frac{1}{2}[\boldsymbol{\sigma}_L^T\ \boldsymbol{\sigma}_g^T]\ [S]\begin{Bmatrix}\boldsymbol{\sigma}_L\\ \boldsymbol{\sigma}_g\end{Bmatrix}dV=\int_V(-\frac{1}{2}\boldsymbol{\varepsilon}_g^T[R_1]\ \boldsymbol{\varepsilon}_g+\frac{1}{2}\boldsymbol{\sigma}_g^T[R_3]\ \boldsymbol{\sigma}_g)\ dV \tag{3-21}$$

and the second term becomes after adding and subtracting $\sigma_L{}^T\ \varepsilon_g$

$$\int_V[\boldsymbol{\sigma}_L^T\ \boldsymbol{\sigma}_g^T]\begin{Bmatrix}\boldsymbol{D}_g\mathbf{u}\\ \boldsymbol{D}_L\mathbf{u}\end{Bmatrix}dV=\int_V(\boldsymbol{\varepsilon}_g^T[R_1]^T\boldsymbol{\varepsilon}_g+\boldsymbol{\sigma}_g^T[R_2]^T\boldsymbol{\varepsilon}_g +\boldsymbol{\sigma}_L^T(\boldsymbol{D}_g\mathbf{u}-\boldsymbol{\varepsilon}_g)+\boldsymbol{\sigma}_g^T\boldsymbol{D}_L\mathbf{u})\ dV \tag{3-22}$$

Substituting them into the functional (3-20), it is modified to

$$\Pi_{III}=\int_V[\frac{1}{2}\boldsymbol{q}^T[R]\ \boldsymbol{q}+\boldsymbol{\sigma}_L^T(\boldsymbol{D}_g\mathbf{u}-\boldsymbol{\varepsilon}_g)+\boldsymbol{\sigma}_g^T\boldsymbol{D}_L\mathbf{u} -\boldsymbol{F}^T\mathbf{u}]\ dV-\int_{S_t}\boldsymbol{T}^T\mathbf{u}\ dS \tag{3-23}$$

If the partial strain-displacement equations (3-8) are satisfied a priori, a new functional is obtained as follows [3.5],

$$\Pi_{CO}=\int_V [E(\mathbf{q})+\boldsymbol{\sigma}_g^T\mathbf{D}_L\mathbf{u}-\mathbf{F}^T\mathbf{u}]\,dV-\int_{S_t}\mathbf{T}^T\mathbf{u}\,dS \tag{3-24}$$

In this new variational functional, there are two fields: displacement field and partial stress field. The constraint conditions are the constitutive equations (3-10), partial strain-displacement equations (3-8), and displacement boundary conditions (3-13). This new variational functional is named by *composite energy functional.* It is different from the Hellinger-Reissner variational functional because the partial strain-displacement equations (3-8) become constraint conditions in the new functional.

Variational Principle of Composite Energy

In order to present variational principle, it is assumed that the composite energy function E is a positive definite function of the components of globally continuous vector, and the body forces and surface forces are derivable from potential functions $\Omega(\mathbf{u})$ and $\Psi(\mathbf{u})$ (2-6). Thus, the principle of composite energy states

Among all the admissible displacement fields and partial stress (transverse stress) fields, which satisfy the partial strain-displacement equations (3-8), constitutive equations (3-10), and prescribed displacement boundary conditions (3-13), the actual displacement field and partial stress field make the total composite energy

$$\Pi_{CO}=\int_V [E(\mathbf{q})+\boldsymbol{\sigma}_g^T\mathbf{D}_L\mathbf{u}-\mathbf{F}^T\mathbf{u}]\,dV-\int_{S_t}\mathbf{T}^T\mathbf{u}\,dS \tag{3-24}$$

an absolute minimum $\delta\Pi_{CO}=0$.

In this principle, the partial strain-displacement equations (3-9), equilibrium equations (3-11), prescribed traction boundary conditions (3-12) are Euler equations. For composites, the conjunction conditions at interlayer surface must be satisfied a priori.

3.3 FORMULATION OF PARTIAL HYBRID ELEMENT

By means of the variational principle of composite energy, one can formulate many new hybrid elements. The term *partial hybrid element* is used to name these new hybrid elements due to the fact that three transverse stresses are only introduced in the assumed stress fields. Within a partial hybrid element, the displacement field is assumed firstly. It is usually described by the nodal displacements as follows,

$$\boldsymbol{u} = [N]\, \boldsymbol{\delta}_e \tag{3-25}$$

where [N] is defined as [N] = [N_1, N_2, ... ,N_n] and [N_i] are vectors of the displacement shape functions. The vector δ_e is nodal displacement vector and n is the number of nodes in the element. The assumed displacement fields in finite elements have been discussed in Chapter 1 and some examples of the displacement shape function [N] will be presented in the next chapter. Then, by means of the partial strain-displacement equations (3-8), the derivatives of the displacements can be expressed as

$$\begin{aligned} \boldsymbol{D}_g\boldsymbol{u} &= [B_g]\, \boldsymbol{\delta}_e = \boldsymbol{\varepsilon}_g \\ \boldsymbol{D}_L\boldsymbol{u} &= [B_L]\, \boldsymbol{\delta}_e \end{aligned} \tag{3-26}$$

Because the partial strain-displacement equations (3-8) are satisfied a priori, the inplane strains can be expressed in the form of the derivatives of the displacements. On the other hand, the transverse strains can not be expressed in that form due to the fact that the partial strain-displacement relations (3-9) are only satisfied a posteriori. In the element, a partial stress field is also assumed independently as follows,

$$\begin{aligned} \boldsymbol{\sigma}_g &= [\ \boldsymbol{\sigma}_1\ \boldsymbol{\sigma}_2\ \ldots\ \boldsymbol{\sigma}_m\ ] \begin{Bmatrix} \beta_1 \\ \beta_2 \\ \cdot \\ \cdot \\ \beta_m \end{Bmatrix} \\ &= [P_g]\, \boldsymbol{\beta} \end{aligned} \tag{3-27}$$

in which, vectors σ_i are stress modes which are functions of the co-ordinates, the parameters β_i are the corresponding stress parameters, and [P] is the stress matrix. Using the definition (3-17), (3-6) and (3-2), the composite energy functional can be written as

$$\Pi_{co}=\int_V [\frac{1}{2}\boldsymbol{\varepsilon}_g^T[R_1]\,\boldsymbol{\varepsilon}_g+\frac{1}{2}\boldsymbol{\sigma}_g^T[R_3]\,\boldsymbol{\sigma}_g+\boldsymbol{\sigma}_g^T[R_2]^T\boldsymbol{\varepsilon}_g$$
$$+\boldsymbol{\sigma}_g^T D_L\mathbf{u}-\boldsymbol{F}^T\mathbf{u}]\,dV-\int_{S_t}\boldsymbol{T}^T\mathbf{u}\,dS \qquad (3\text{-}28)$$

Substituting the equations (3-25)-(3-27) into the equation (3-28), the functional becomes

$$\begin{aligned}\Pi_{co}=&\frac{1}{2}\boldsymbol{\delta}_e^T\int_V [B_g]^T[R_1]\,[B_g]\,dV\,\boldsymbol{\delta}_e\\&+\frac{1}{2}\boldsymbol{\beta}^T\int_V [P_g]^T[R_3]\,[P_g]\,dV\,\boldsymbol{\beta}\\&+\boldsymbol{\beta}^T\int_V [P_g]^T([B_L]+[R_2]^T[B_g])\,dV\,\boldsymbol{\delta}_e\\&-\boldsymbol{\delta}_e^T\int_V [N]^T\boldsymbol{F}dV\;-\boldsymbol{\delta}_e^T\int_{S_t}[N]^T\boldsymbol{T}dS\end{aligned} \qquad (3\text{-}29)$$

Denote

$$\begin{aligned}&[H]=-\textstyle\int_V[P_g]^T[R_3]\,[P_g]\,dV\\&[G]=\textstyle\int_V[P_g]^T([B_L]+[R_2]^T[B_g])\,dV\\&[K_d]=\textstyle\int_V[B_g]^T[R_1]\,[B_g]\,dV\\&\boldsymbol{f}=\int_V[N]^T\boldsymbol{F}dV+\int_{S_t}[N]^T\boldsymbol{T}dS\end{aligned} \qquad (3\text{-}30)$$

Thus, the functional is expressed as

$$\Pi_{co}=\frac{1}{2}\boldsymbol{\delta}_e^T[K_d]\,\boldsymbol{\delta}_e-\frac{1}{2}\boldsymbol{\beta}^T[H]\,\boldsymbol{\beta}+\boldsymbol{\beta}^T[G]\,\boldsymbol{\delta}_e-\boldsymbol{\delta}_e^T\boldsymbol{f} \qquad (3\text{-}31)$$

In this variational functional, there are two independent variables subject to variation. From the partial stationary condition with respect to β,

$$\frac{\partial \Pi_{co}}{\partial \boldsymbol{\beta}}=0 \tag{3-32}$$

the relation between stress parameters β and nodal displacements δ_e is obtained,

$$[H]\,\boldsymbol{\beta}=[G]\,\boldsymbol{\delta}_e \tag{3-33}$$

By means of this relation, then, the functional (3-31) becomes

$$\Pi_{co}=\frac{1}{2}\boldsymbol{\delta}_e^T[K_d]\,\boldsymbol{\delta}_e+\frac{1}{2}\boldsymbol{\delta}_e^T(\,[G]^T[H]^{-1}[G]\,)\,\boldsymbol{\delta}_e-\boldsymbol{\delta}_e^T\boldsymbol{f}_e \tag{3-34}$$

Denote

$$[K_h]=[G]^T[H]^{-1}[G] \qquad \textit{and} \qquad [K]_e=[K_d]+[K_h] \tag{3-35}$$

Then, equation (3-34) can be rewritten as

$$\Pi_{co}=\frac{1}{2}\boldsymbol{\delta}_e^T[K]_e\boldsymbol{\delta}_e-\boldsymbol{\delta}_e^T\boldsymbol{f}_e \tag{3-36}$$

From the partial stationary condition with respect to δ_e, the governing equation of the element is obtained,

$$[K]_e\,\boldsymbol{\delta}_e=\boldsymbol{f}_e \tag{3-37}$$

in which, $[K]_e$ is the element stiffness matrix. For the partial hybrid element, the element stiffness matrix consists of a displacement-formulated stiffness matrix $[K_d]$ and a hybrid-formulated stiffness matrix $[K_h]$.

3.4 DETERMINATION OF STRESS MODES

For hybrid stress elements, there have been problems with the determination of the stress polynomials. While the displacement polynomials are constrained by the number of displacement nodal degrees of freedom in the element, the stress polynomials have no such constraint. However, if not enough polynomial terms are introduced into the stress field, then there may exist spurious kinematic deformation modes in the stiffness matrix which make it singular. One possible way to remedy this problem is to add more terms into the stress polynomials. But there is no guarantee that the additional terms will resolve the singularity problem. Also, the more terms there are in the stress polynomials, the larger the matrices [H] and [G] will become and the more computer resources will be required. In addition, if there are too many terms in the stress polynomials, the resulting element model will be over-rigid. Furthermore, overuse of the terms in the stress polynomials will cause locking in some elements such as plate/shell elements. Although there are several approaches to determine the optimal assumed stress field [3.20-3.28], the problem has been not solved completely. In order to solve these problems, this section presents the eigenfunction method [3.5], the iso-function method [3.7], and the classification method [3.16] for determination of stress polynomials. For simplicity, the discussion is restricted to the conventional hybrid elements based on the Hellinger-Reissner variational principle. However, these techniques can also be applied to partial hybrid elements. Once the displacement polynomials for an element are determined, by means of these techniques, the stress polynomials will be constrained by optimal condition of the stress field.

3.4.1 Eigenfunction Method

A finite element has a finite number of degrees of freedom. For instance, a 2-D, 4-node displacement element has (n=) 8 degrees of freedom, and a 3-D, 8-node displacement element has (n=) 24 degrees of freedom. Therefore, the displacement field of an element can be described by n nodal displacements. It can also be described by n-r independent deformation modes and r rigid body modes. Thus, it can be assumed that there exist m (=n-r) natural deformation modes and r rigid body modes in the element. The displacement distribution in the element can be represented by them.

In finite element method, for both single-field displacement formulation and multifield hybrid formulation, the governing equation of nodal displacements has the same form as follows,

$$[K]_e \boldsymbol{\delta}_e = \boldsymbol{f}_e \tag{3-38}$$

If the nodal force vector is proportional to the nodal displacement vector, the equilibrium equation (3-38) becomes a eigenvalue equation. It can be expressed as follows

$$([K]_e - \lambda [I]) \boldsymbol{\delta}_e = 0 \tag{3-39}$$

where $[K]_e$ is a nxn element stiffness matrix. This equation will give (n-r) non-zero eigenvalues and r zero eigenvalues, and (n-r) eigenvectors corresponding to the (n-r) non-zero eigenvalues. If vectors $\{\delta_i\}$ (i=1,2,...m) are the eigenvectors of the stiffness matrix $[K]_e$, they must satisfy the orthogonal condition:

$$\begin{aligned} \{\boldsymbol{\delta}_i\}^T\{\boldsymbol{\delta}_j\}&=0 \qquad i\neq j \\ \{\boldsymbol{\delta}_i\}^T\{\boldsymbol{\delta}_j\}&=1 \qquad i=j \end{aligned} \tag{3-40}$$

Natural Deformation Mode

In the single-field displacement element, the stiffness matrix can be expressed in the form (see chapter 1),

$$[K]_e = \int_{V_e} [B]^T [C] [B] \, dV \tag{1-6}$$

In which, [C] is the material constant matrix and [B] is the geometry matrix of the element. Therefore, the eigenvalues and eigenvectors only depend on the material properties and geometry of the element. Due to the fact that the (n-r) eigenvectors $\{\delta\}_i$ (i=1,2,3, ... m) are unique, they can be considered as the *natural deformation modes* of the element with a special shape and material constants [3.5, 3.29].

These natural deformation modes are independent from each other and are the basic deformation modes of the element. Any deformation in the element can be described by the linear combination of these basic deformation modes and the energy of the element is decomposable into these orthogonal modes.

Natural Stress Mode

In the hybrid element based on the Hellinger-Reissner variational principle, the stiffness matrix can be expressed in the form (see chapter 2),

$$[K]_e = [G]^T [H]^{-1} [G]$$

$$[H] = \int_{V_e} [P]^T [S] [P] \, dV \tag{2-28}$$

$$[G] = \int_{V_e} [P]^T [B] \, dV$$

where [H] is the flexibility matrix, [G] is the leverage matrix, [S] is the material constant matrix, and [P] is the stress matrix. Because of the existence of the matrix [P], the eigenvalues and eigenvectors of the element will not only depend on the material properties and the geometric shape of the element, but also be sensitive to the stress modes in the assumed stress field.

In hybrid elements, an assumed stress matrix [P] may contain zero-energy stress modes and its resulting stiffness matrix may have spurious kinematic deformation modes. The zero-energy stress modes are such stress modes that do not produce deformation energy. The eigenvalues of the element stiffness matrix corresponding to these stress modes equal zero, and these stress modes correspond to rigid body modes. The kinematic deformation modes are these deformation modes corresponding to spurious zero stiffness. They may be caused by unsuitable numerical integration technique or unsuitable assumed stress fields. In this book, the numerical integration technique is not discussed. The selection of stress modes will be only discussed here. Therefore, non-zero-energy tress modes should correspond to the natural deformation modes [3.5,3.29]. Thus, it can be stated that:

There are m (=n-r) *natural stress modes* in a hybrid element with n degrees of freedom and r rigid body modes. These natural stress modes correspond to m natural deformation modes which are orthogonal and independent from each other, and the energy of the hybrid element is decomposable into these orthogonal modes.

Postulate 1

If and only if the elastic energy of the structure with finite degrees of freedom is decomposable, the eigenvalues obtained from separate stress mode equations

$$([K_i]_e - \lambda [I]) \delta_e = 0 \tag{3-41}$$

are equal to the eigenvalues obtained from the total stress mode equation

$$([K]_e - \lambda [I])\,\delta_e = 0 \tag{3-42}$$

where i=1,2, ... , m. The stiffness matrix [K] is defined by equations (2-28) and (2-34). It is

$$\begin{aligned} [K]_e &= [G]^T [H]^{-1} [G] \\ [H] &= \int_{V_e} [P]^T [S]\, [P]\, dV \\ [G] &= \int_{V_e} [P]^T [B]\, dV \end{aligned} \tag{2-34)'$$

The matrix $[K_i]$ is defined as follows,

$$\begin{aligned} [K_i]_e &= [G_i]^T [H_i]^{-1} [G_i] \\ [H_i] &= \int_{V_e} \{\sigma_i\}^T [S]\, \{\sigma_i\} dV \\ [G_i] &= \int_{V_e} \{\sigma_i\}^T [B]\, dV \end{aligned} \tag{3-43}$$

in which

$$[P] = [\{\sigma_1\}\ \{\sigma_2\}\ \{\sigma_3\}\ \ldots\ldots\ \{\sigma_m\}\ ]$$

$$[G] = \begin{Bmatrix} [G_1] \\ [G_2] \\ \cdots \\ [G_m] \end{Bmatrix} \tag{3-44}$$

This postulate [3.5,3.29] can be stated as a theorem:

Theorem 1

For a hybrid element with n degrees of freedom and r rigid body modes, if and only if the matrix [H] is a diagonal matrix, the stiffness matrix satisfies the superposition principle:

$$[K]_e = \sum_{i=1}^{m} [K_i]_e \tag{3-45}$$

where the matrices $[K]_e$ and $[K_i]_e$ are defined in the equations (2-34)' and (3-43).

Proof [3.5]. Assume [H] to be a diagonal matrix and denote:

$$c_i = \frac{1}{H_{ii}} \tag{3-46}$$

The inversion of the diagonal matrix [H] can be written in the form,

$$[H]^{-1} = \begin{bmatrix} c_1 & 0 & \dots & \dots & 0 \\ 0 & c_2 & \dots & \dots & 0 \\ . & & . & & . \\ . & & & . & . \\ 0 & . & . & . & c_m \end{bmatrix} \tag{3-47}$$

and

$$[H_i]^{-1} = [c_i]$$

Thus, by means of equations (2-34)' and (3-44), the element stiffness matrix is expressed as follows,

$$[K]_e = [\ [G_1]^T\ [G_2]^T \dots [G_m]^T\ ] \begin{bmatrix} c_1 & 0 & \dots & \dots & 0 \\ 0 & c_2 & \dots & \dots & 0 \\ . & & . & & . \\ . & & & . & . \\ 0 & . & . & . & c_m \end{bmatrix} \begin{Bmatrix} [G_1] \\ [G_2] \\ . \\ . \\ [G_m] \end{Bmatrix} \tag{3-48}$$

Then,

$$[K]_e=\sum_{1}^{m}[G_1]^T[c_1][G_1]$$
$$=\sum_{1}^{m}[G_1]^T[H_1]^{-1}[G_1] \qquad (3\text{-}49)$$
$$=\sum_{1}^{m}[K_1]_e$$

■

Due to the fact that the matrix [H] is a diagonal matrix if the related stress matrix [P] consists of the natural stress modes of an element, equation (3-49) shows that the natural modes are independent from each other and the elastic energy of the element is decomposable into these natural modes.

Determination of Natural Stress Modes

In order to get the natural stress modes from the natural deformation modes, the relation (2-31) between stress parameters β and nodal displacements δ_e is used. It is

$$[H]\,\boldsymbol{\beta}=[G]\,\boldsymbol{\delta}_e \qquad (2\text{-}31)$$

For simplicity, the subscript e of the nodal displacement vector in the equation (2-31) will be omitted in the following text. If the vectors δ_l (l=1,2,...,m) are the eigenvectors of the stiffness matrix of the single-field displacement element, they are considered as the natural deformation modes of its hybrid counterpart because they do not depend on the stress modes. Thus, in eigenfunction method, a set of initial stress modes can be assumed. Subsequently, these initial stress modes can be modified by means of iterative process as follows

$$\boldsymbol{\beta}^i=[H^i]^{-1}[G^i]\,\boldsymbol{\delta}_1$$
$$\boldsymbol{\sigma}_1^{i+1}=[P^i]\,\boldsymbol{\beta}^i \qquad (3\text{-}50)$$

in which, superscript i represents the i-th step of the iterative process. Details of the iterative process are shown in the following pages. This process continues until the matrix [H] becomes diagonal and the stress matrix [P] is stationary. The resulting stress modes from this process are considered as the natural stress modes corresponding to the natural deformation modes. Theoretically, the natural stress modes of an element can be found by using the eigenfunction method. However, in the case of multiple eigenvalues, modification of the eigenvectors has to be made in numerical iterative process.

The procedure of the eigenfunction method is as follows [3.5, 3.29],

Step 1. Calculate the eigenvectors of the element stiffness matrix of the displacement element having the same displacement field as that by a hybrid element:

$$\boldsymbol{\delta}_l \qquad l=1,2,\ldots,m \tag{3-51}$$

Step 2. Assume a complete set of the stress field modes as follows,

$$[P^0] = [\boldsymbol{\sigma}_1^0 \ \boldsymbol{\sigma}_2^0 \ \ldots \ \boldsymbol{\sigma}_L^0] \tag{3-52}$$

where L >= m (=n-r).

Step 3. For i = 1,2,3,...,k, where k is the number of iteration, run the iterative steps 4 - 6 until the matrix [H] becomes diagonal and the stress matrix [P] is stationary.

Step 4. Calculate the matrices

$$[H^i] = \int_{V_e} [P^i]^T [S] [P^i] \, dV \tag{3-53}$$
$$[G^i] = \int_{V_e} [P^i]^T [B] \, dV$$

Step 5. Modify m stress modes in the matrix $[P^i]$

$$\boldsymbol{\sigma}_l^{i+1} = [P^i] [H^i]^{-1} [G^i] \boldsymbol{\delta}_l \tag{3-54}$$

where l=1,2,...,m.

Step 6. Normalize the stress modes

$$\mu_i^{i+1} = \sigma_i^{i+1} / \sqrt{\sigma_i^{i+1} \cdot \sigma_i^{i+1}}$$

$$\sigma_i^{i+1} = \frac{\mu_i^{i+1}}{max(|\mu_1^{i+1}|, |\mu_2^{i+1}|, \ldots, |\mu_m^{i+1}|)} \tag{3-55}$$

$$[P^{i+1}] = [\sigma_1^{i+1} \ \sigma_2^{i+1} \ \ldots \ \sigma_m^{i+1}]$$

If the matrix [H] becomes diagonal, the stress modes in the stress matrix [P] are the natural stress modes of the hybrid element. One caution needs to be considered. In the eigenvalue analysis, there may exist multiple eigenvalues. In this case, there are a lot of choices for the directions of the corresponding eigenvectors. However, it is not guaranteed that every choice can result in a stress mode which gives a diagonal matrix [H] and makes element stiffness matrix [K] superpositionable. Therefore, for multiple eigenvalues, the orthogonal displacement eigenvectors may or may not result in orthogonal stress modes.

Examples of Stress Matrix Determined by Eigenfunction Method

2-D, 4-node Hybrid Element

In the hybrid element with isotropic elastic material and rectangular shape, the number of the degrees of freedom is eight (n=8), and the number of rigid body modes is three (r=3). Therefore, the number of stress modes in the assumed stress field must be larger than five (m=n-r=5). For two different sets of initial stress modes, the stress matrices [P^i] in the iterative process are given as follows [3.5,3.29],

Example 1:

Step 1. Calculate the eigenvectors (δ_1 , δ_2 , δ_3 , δ_4, δ_5) of the stiffness matrix [K] of the displacement element.

Step 2. Assume a stress matrix,

$$[P^0] = \begin{bmatrix} 1 & 1 & 0 & \eta & 0 & 0 & -\xi \\ 1 & -1 & 0 & 0 & \xi & -\eta & 0 \\ 0 & 0 & 1 & 0 & 0 & \xi & \eta \end{bmatrix} \tag{3-56}$$

Step 3. Run the iterative steps 4 - 6 for i = 1,2,3, ...

$$[P^1] = \begin{bmatrix} -\eta & 0.045\xi & 0 & 1 & -1 \\ 0.045\eta & -\xi & 0 & -1 & -1 \\ -0.045\xi & -0.045\eta & 1 & 0 & 0 \end{bmatrix} \tag{3-57}$$

$$[P^2] = \begin{bmatrix} -\eta & 0.002\xi & 0 & 1 & -1 \\ 0.002\eta & -\xi & 0 & -1 & -1 \\ -0.002\xi & -0.002\eta & 1 & 0 & 0 \end{bmatrix} \tag{3-58}$$

$$[P^3] = \begin{bmatrix} -\eta & 4\times10^{-6}\xi & 0 & 1 & -1 \\ 4\times10^{-6}\eta & -\xi & 0 & -1 & -1 \\ -4\times10^{-6}\xi & -4\times10^{-6}\eta & 1 & 0 & 0 \end{bmatrix} \tag{3-59}$$

$$[P^4] = \begin{bmatrix} -\eta & 2\times10^{-11}\xi & 0 & 1 & -1 \\ 2\times10^{-11}\eta & -\xi & 0 & -1 & -1 \\ -2\times10^{-11}\xi & -2\times10^{-11}\eta & 1 & 0 & 0 \end{bmatrix} \tag{3-60}$$

$$[P^5] = \begin{bmatrix} -\eta & 0 & 0 & 1 & -1 \\ 0 & -\xi & 0 & -1 & -1 \\ 0 & 0 & 1 & 0 & 0 \end{bmatrix} \tag{3-61}$$

When i=5, the matrix [H] becomes diagonal. Therefore, the final stress matrix is obtained.

Example 2:

Step 1. Calculate the eigenvectors (δ_1 , δ_2 , δ_3 , δ_4, δ_5) of the stiffness matrix [K] of the displacement element.

Step 2. Assume a stress matrix,

$$[P^0] = \begin{bmatrix} 1 & 1 & 0 & \xi & 0 & 0 & \eta & 0 & 0 \\ 1 & -1 & 0 & 0 & \xi & 0 & 0 & \eta & 0 \\ 0 & 0 & 1 & 0 & 0 & \xi & 0 & 0 & \eta \end{bmatrix} \tag{3-62}$$

Step 3. Run the iterative steps 4 - 6 for i = 1,2,3, ...

$$[P^1] = \begin{bmatrix} -\eta & 0.09\xi & 0 & 1 & -1 \\ 0.09\eta & -\xi & 0 & -1 & -1 \\ -0.12\xi & -0.12\eta & 1 & 0 & 0 \end{bmatrix} \tag{3-63}$$

$$[P^2] = \begin{bmatrix} -\eta & -0.008\xi & 0 & 1 & -1 \\ -0.008\eta & -\xi & 0 & -1 & -1 \\ -0.015\xi & -0.015\eta & 1 & 0 & 0 \end{bmatrix} \tag{3-64}$$

$$[P^3] = \begin{bmatrix} -\eta & -6\times10^{-5}\xi & 0 & 1 & -1 \\ -6\times10^{-5}\eta & -\xi & 0 & -1 & -1 \\ -2\times10^{-4}\xi & -2\times10^{-4}\eta & 1 & 0 & 0 \end{bmatrix} \tag{3-65}$$

$$[P^4] = \begin{bmatrix} -\eta & -4\times10^{-9}\xi & 0 & 1 & -1 \\ -4\times10^{-9}\eta & -\xi & 0 & -1 & -1 \\ -5\times10^{-8}\xi & -5\times10^{-8}\eta & 1 & 0 & 0 \end{bmatrix} \tag{3-66}$$

......

......

$$[P^7] = \begin{bmatrix} -\eta & 0 & 0 & 1 & -1 \\ 0 & -\xi & 0 & -1 & -1 \\ 0 & 0 & 1 & 0 & 0 \end{bmatrix} \tag{3-67}$$

When i=7, the matrix [H] becomes diagonal. Therefore, the final stress matrix is obtained.

It can be seen that the final stress matrices [P] are the same although they start from the different initial stress modes.

3-D, 8-node Hybrid Element

In 3-D, 8-node hybrid solid element with isotropic elastic material, the number of the degrees of freedom equals twenty four (n=24), and the number of rigid body modes equals six (r=6). Therefore, the number of stress modes in the assumed stress field must be larger than eighteen (m=n-r=18). Similar to the 2-D case above, the natural stress modes are obtained by Huang [3.5,3.29]. They are

Tension and compressive modes,

$$\{\sigma_1\}=\begin{Bmatrix} 1 \\ 1 \\ 1 \\ 0 \\ 0 \\ 0 \end{Bmatrix} \quad \{\sigma_2\}=\begin{Bmatrix} 1 \\ -1 \\ 0 \\ 0 \\ 0 \\ 0 \end{Bmatrix} \quad \{\sigma_3\}=\begin{Bmatrix} -1 \\ -1 \\ 2 \\ 0 \\ 0 \\ 0 \end{Bmatrix} \tag{3-68}$$

Pure shear modes,

$$\{\sigma_4\}=\begin{Bmatrix} 0 \\ 0 \\ 0 \\ 1 \\ 0 \\ 0 \end{Bmatrix} \quad \{\sigma_5\}=\begin{Bmatrix} 0 \\ 0 \\ 0 \\ 0 \\ 1 \\ 0 \end{Bmatrix} \quad \{\sigma_6\}=\begin{Bmatrix} 0 \\ 0 \\ 0 \\ 0 \\ 0 \\ 1 \end{Bmatrix} \tag{3-69}$$

Symmetric bending modes,

$$\{\sigma_7\}=\begin{Bmatrix} \zeta \\ \zeta \\ 0 \\ 0 \\ 0 \\ 0 \end{Bmatrix} \quad \{\sigma_8\}=\begin{Bmatrix} 0 \\ \xi \\ \xi \\ 0 \\ 0 \\ 0 \end{Bmatrix} \quad \{\sigma_9\}=\begin{Bmatrix} \eta \\ 0 \\ \eta \\ 0 \\ 0 \\ 0 \end{Bmatrix} \tag{3-70}$$

Anti-symmetric bending modes,

$$\{\sigma_{10}\}=\begin{Bmatrix} \zeta \\ -\zeta \\ 0 \\ 0 \\ 0 \\ 0 \end{Bmatrix} \quad \{\sigma_{11}\}=\begin{Bmatrix} 0 \\ \xi \\ -\xi \\ 0 \\ 0 \\ 0 \end{Bmatrix} \quad \{\sigma_{12}\}=\begin{Bmatrix} \eta \\ 0 \\ -\eta \\ 0 \\ 0 \\ 0 \end{Bmatrix} \tag{3-71}$$

Torsion modes,

$$\{\sigma_{13}\}=\begin{Bmatrix} 0 \\ 0 \\ 0 \\ \zeta \\ \xi \\ \eta \end{Bmatrix} \quad \{\sigma_{14}\}=\begin{Bmatrix} 0 \\ 0 \\ 0 \\ \zeta \\ -\xi \\ 0 \end{Bmatrix} \quad \{\sigma_{15}\}=\begin{Bmatrix} 0 \\ 0 \\ 0 \\ \zeta \\ \xi \\ -2\eta \end{Bmatrix} \tag{3-72}$$

Saddle distributed modes,

$$\{\sigma_{16}\}=\begin{Bmatrix} \eta\zeta \\ 0 \\ 0 \\ 0 \\ 0 \\ 0 \end{Bmatrix} \quad \{\sigma_{17}\}=\begin{Bmatrix} 0 \\ \zeta\xi \\ 0 \\ 0 \\ 0 \\ 0 \end{Bmatrix} \quad \{\sigma_{18}\}=\begin{Bmatrix} 0 \\ 0 \\ \xi\eta \\ 0 \\ 0 \\ 0 \end{Bmatrix} \tag{3-73}$$

These stress modes can formulate a diagonal matrix [H] for the hybrid element, and resulting stiffness matrix [K] satisfies the superposition principle. Therefore, the energy of the hybrid element is decomposable. Moreover, the eigenvalues of the stiffness matrix [K] (2-34)' is equal to that of the stiffness matrix $[K_i]$ (3-43).

3.4.2 Iso-function Method

The eigenfunction method can not be used to derive partial stress field for partial hybrid element due to the existence of multiple eigenvalues. In order to assume a partial stress field without zero energy modes, a new method termed *iso-function method* is presented [3.6-3.7]. The method presents an easy way to form a stress field although the formulated stress field may not be better than others [3.21-

3.28].

Iso-function Stress Matrix

Within a hybrid finite element based on the Hellinger-Reissner variational principle, a displacement field is assumed in the form,

$$\mathbf{u} = [\Phi]\,\mathbf{a} \tag{3-74}$$

where $[\Phi]$ is defined as $[\Phi] = [\Phi_1\ \Phi_2\ \ldots\ \Phi_n]$, and $\{\Phi_i\}$ are vectors of displacement interpolation functions. They only depend on the element geometry and have no relation with the material properties of the element. $\mathbf{a} = [a_1\ a_2\ \ldots\ a_n]^T$ is the displacement parameter vector, and the subscript n is the total number of degrees of freedom of the element. By means of strain-displacement relation (2-1), then, the strain field is derived from the displacement field as follows,

$$\varepsilon = [D]\,\mathbf{u} = [\Psi]\,\mathbf{a} \tag{3-75}$$

where $[D]$ is the derivative operator matrix and $[\Psi] = [D][\Phi]$. The displacement-derived stress field can be obtained using the constitutive relation (2-2),

$$\bar{\boldsymbol{\sigma}} = [C]\,\boldsymbol{\varepsilon} = [C]\,[\boldsymbol{\Psi}]\,\boldsymbol{a} \tag{3-76}$$

The expression can be rewritten in the form

$$\bar{\boldsymbol{\sigma}} = [\boldsymbol{\Theta}]\,\boldsymbol{\gamma} \tag{3-77}$$

where $[\Theta]$ is a function of co-ordinates and γ is a vector of stress parameters which are the products of material constants C_{ij} and displacement parameters a_k. Therefore, the stress parameters depend on the geometry, material properties, and displacement parameters of the element. On the other hand, the stress field in the conventional hybrid element is independently assumed as

$$\boldsymbol{\sigma} = [P]\,\boldsymbol{\beta} \tag{3-78}$$

where $[P]$ is a stress matrix which is functions of co-ordinates, and $\boldsymbol{\beta}$ is a vector of stress parameters which depends on the geometry, material properties, and displacement parameters of the element. If the number of stress parameters in the two stress fields (3-77) and (3-78) is same, the difference between them is only that the stress parameters (3-77) are not basic variables and depend on the displacement variables, whereas the stress parameters (3-78) are basic variables. However, the two stress fields represent the same field in an element. If the assumed displacement field (3-74) does exactly describe the deformation of the element and the assumed stress field (3-78) does exactly represent the stress distribution in the element, in the case of n $\rightarrow\infty$, the displacement-derived stress field and the assumed stress field must be same. That is

$$[P]\,\boldsymbol{\beta} = [\boldsymbol{\Theta}]\,\boldsymbol{\gamma} \tag{3-79}$$

Therefore, without loss of generality, it is assumed that

$$[\,P\,] = [\,\Theta\,] \tag{3-80}$$

In this relation, the stress matrix [P] is assumed to be the same as the displacement-derived stress matrix. Thus, the matrix [P] is called *iso-function stress matrix* and the method to establish an assumed stress field is called *iso-function method.*

Iso-function Partial Stress Matrix

For a partial hybrid element based on the variational principle of composite energy, the displacement-derived stress field (3-77) can be split to two parts as follows,

$$\begin{Bmatrix} \boldsymbol{\sigma}_g \\ \boldsymbol{\sigma}_L \end{Bmatrix} = \begin{bmatrix} \boldsymbol{\Theta}_g \\ \boldsymbol{\Theta}_L \end{bmatrix} \boldsymbol{\gamma} \tag{3-81}$$

and a partial stress field is independently assumed in the form,

$$\boldsymbol{\sigma}_g = [P_g]\,\boldsymbol{\beta} \tag{3-82}$$

Similarly, in the case of n$\rightarrow\infty$, the assumed partial stress field (3-82) should be equivalent to the displacement-derived partial stress field (3-81) if they do represent the real stress distribution for the same element. Therefore, it is assumed that the assumed partial stress matrix is equal to the displacement-derived partial stress matrix. That is

$$[P_g] = [\boldsymbol{\Theta}_g] \tag{3-83}$$

Using this various partial stress fields can be derived for the different partial hybrid elements. However, one question remains. That is the possibility that there are zero-energy stress modes which may cause spurious kinematic deformation modes in the element. Before answering this question, the relationship between the convention displacement element and hybrid element constructed by iso-function stress matrix should be discussed.

Equivalence Between Hybrid Element and Displacement Element

In the process of deriving iso-function stress matrix, it is assumed that n $\rightarrow\infty$. In practice, n is a finite number. The assumed displacement field is only an approximation of the actual deformation state of an element. Thus, the problem becomes: Is it possible that an assumed stress field for a hybrid element is exactly equivalent to its displacement-derived stress field when n is a finite number ? The answer is YES. The proof is as follows.

In Chapter 2, it has been indicated that there are the limitation principles [3.30-3.31] which establish the relationship between hybrid elements and displacement elements. For hybrid elements based on the Hellinger-Reissner variational principle, it states

A hybrid element would be equivalent to its displacement counterpart if the displacement-derived stress space is a subspace of the assumed stress.

This means that a hybrid element based on the Hellinger-Reissner variational principle would be no different from a displacement element when the assumed stress field contains all stress modes derived from the assumed displacement field. Due to the fact that the iso-function stress matrix is directly derived from the displacement field, the assumed stress field constructed by iso-function stress matrix will absolutely contain all stress modes which can be derived from the displacement field. Therefore, the hybrid element constructed by iso-function method is equivalent to its displacement counterpart. As a result, the assumed stress field is the same as the displacement-derived stress field. Because a displacement element does not have any kinematic deformation mode, the equivalent hybrid element does not have any kinematic deformation mode and assumed stress field does not contain any zero-energy mode.

For the partial hybrid element, the works on the limitation principle were extended to the partial hybrid plate/shell element [3.32,3.14]. Similarly, one also can extend the limitation principles to the partial hybrid elements based on the variational principle of composite energy.

Limitation principle. A partial hybrid element would be equivalent to its displacement counterpart if the displacement-derived partial stress space is a subspace of the assumed partial stress.

By means of this limitation principle, it is clear that the partial hybrid element using iso-function method is equivalent to its displacement counterpart because the iso-function partial stress matrix contains all displacement-derived stress modes. Therefore, similar to the conventional hybrid element, the assumed partial stress field is equivalent to the displacement-derived partial stress field and is free from any zero-energy mode.

Examples of Iso-function Stress Matrix

Iso-function method can be used to establish the assumed stress fields and assumed partial stress fields for hybrid elements and partial hybrid elements. Here are some examples.

1 Iso-function Stress Matrix for 2-D, 3-node Element

Within this element, there are six degrees of freedom and the assumed displacement field has six parameters. It can be expressed in the form,

$$\begin{aligned} u &= a_0 + a_1\xi + a_2\eta \\ v &= b_0 + b_1\xi + b_2\eta \end{aligned} \tag{3-84}$$

Using strain-displacement relation, the strains are derived as follows,

$$\begin{aligned} \varepsilon_\xi &= \frac{\partial u}{\partial \xi} = a_1 \\ \varepsilon_\eta &= \frac{\partial v}{\partial \eta} = b_2 \\ \varepsilon_{\xi\eta} &= \frac{\partial u}{\partial \eta} + \frac{\partial v}{\partial \xi} = a_2 + b_1 \end{aligned} \tag{3-85}$$

For linear elastic body, the material stiffness matrix is

$$[C] = \begin{bmatrix} C_{11} & C_{12} & C_{13} \\ C_{21} & C_{22} & C_{23} \\ C_{31} & C_{32} & C_{33} \end{bmatrix} \tag{3-86}$$

Using the constitutive relation (3-76), the displacement-derived stress field is

$$\overline{\sigma_\xi}=a_1c_{11}+b_2c_{12}+(a_2+b_1)\,c_{13}=\gamma_1$$
$$\overline{\sigma_\eta}=a_1c_{21}+b_2c_{22}+(a_2+b_1)\,c_{23}=\gamma_2 \tag{3-87}$$
$$\overline{\sigma_{\xi\eta}}=a_1c_{31}+b_2c_{32}+(a_2+b_1)\,c_{33}=\gamma_3$$

It can be written in the matrix form (3-77),

$$\overline{\boldsymbol{\sigma}}=[\boldsymbol{\Theta}]\,\boldsymbol{\gamma}=\begin{bmatrix}1&0&0\\0&1&0\\0&0&1\end{bmatrix}\boldsymbol{\gamma} \tag{3-88}$$

where [Θ] is the function of co-ordinates and γ is the vector of the stress parameters γ_i which are the products of material constants C_{ij} and displacement parameters a_k. Therefore, the stress parameters depend on the geometry, material properties, and displacement parameters of the element. By means of iso-function relation

$$[\,P\,]=[\,\Theta\,] \tag{3 - 80}$$

The stress matrix [P] is obtained. It is

$$[P]=\begin{bmatrix}1&0&0\\0&1&0\\0&0&1\end{bmatrix} \tag{3-89}$$

Therefore, the assumed stress field is

$$\boldsymbol{\sigma}=[P]\,\boldsymbol{\beta}=\begin{bmatrix}1&0&0\\0&1&0\\0&0&1\end{bmatrix}\boldsymbol{\beta} \tag{3-90}$$

in which, stress parameters are basic variables in variational functional and are determined by the variational principle.

2 Iso-function Stress Matrix for 2-D, 4-node Element

For this element, the number of degrees of freedom is eight and assumed displacement field has eight parameters. It can be expressed in the form,

$$u=a_0+a_1\xi+a_2\eta+a_3\xi\eta$$
$$v=b_0+b_1\xi+b_2\eta+b_3\xi\eta \tag{3-91}$$

Using strain-displacement relation, the strains are derived as follows,

$$\varepsilon_\xi=\frac{\partial u}{\partial \xi}=a_1+a_3\eta$$
$$\varepsilon_\eta=\frac{\partial v}{\partial \eta}=b_2+b_3\xi \tag{3-92}$$
$$\varepsilon_{\xi\eta}=\frac{\partial u}{\partial \eta}+\frac{\partial v}{\partial \xi}=a_2+b_1+a_3\xi+b_3\eta$$

Using the constitutive relation (3-76) and the stiffness matrix (3-86), the displacement-derived stress field is

$$\bar{\boldsymbol{\sigma}}=[\boldsymbol{\Theta}]\,\boldsymbol{\gamma} \tag{3-93}$$

In which,

$$[\boldsymbol{\Theta}]=\begin{bmatrix} 1 & 0 & 0 & \xi & 0 & 0 & \eta & 0 & 0 \\ 0 & 1 & 0 & 0 & \xi & 0 & 0 & \eta & 0 \\ 0 & 0 & 1 & 0 & 0 & \xi & 0 & 0 & \eta \end{bmatrix} \tag{3-94}$$

and

$$\begin{aligned}
\gamma_1 &= c_{11}a_1+c_{12}b_2+c_{13}a_2+c_{13}b_1 \\
\gamma_4 &= c_{12}b_3+c_{13}a_3 \\
\gamma_7 &= c_{11}a_3+c_{13}b_3 \\
\gamma_2 &= c_{21}a_1+c_{22}b_2+c_{23}a_2+c_{23}b_1 \\
\gamma_5 &= c_{22}b_3+c_{23}a_3 \\
\gamma_8 &= c_{21}a_3+c_{23}b_3 \\
\gamma_3 &= c_{31}a_1+c_{32}b_2+c_{33}a_2+c_{33}b_1 \\
\gamma_6 &= c_{32}b_3+c_{33}a_3 \\
\gamma_9 &= c_{31}a_3+c_{33}b_3
\end{aligned} \tag{3-95}$$

In which, [Θ] is the function of co-ordinates and γ is the vector of the stress parameters which are the products of material constants C_{ij} and displacement parameters a_k. By means of iso-function relation (3-80), the stress matrix [P] is obtained and the assumed stress field is

$$\boldsymbol{\sigma}=[P]\,\boldsymbol{\beta}=\begin{bmatrix} 1 & 0 & 0 & \xi & 0 & 0 & \eta & 0 & 0 \\ 0 & 1 & 0 & 0 & \xi & 0 & 0 & \eta & 0 \\ 0 & 0 & 1 & 0 & 0 & \xi & 0 & 0 & \eta \end{bmatrix}\boldsymbol{\beta} \tag{3-96}$$

In which, stress parameters are basic variables in variational functional and are determined by the variational principle.

3 Iso-function Stress Matrix for 3-D, 8-node Element

For a 3-D, 8-node hybrid element, the displacement field is assumed in the form,

$$\begin{aligned}
u &= a_0+a_1\xi+a_2\eta+a_3\zeta+a_4\xi\eta+a_5\xi\zeta+a_6\eta\zeta+a_7\xi\eta\zeta \\
v &= b_0+b_1\xi+b_2\eta+b_3\zeta+b_4\xi\eta+b_5\xi\zeta+b_6\eta\zeta+b_7\xi\eta\zeta \\
v &= c_0+c_1\xi+c_2\eta+c_3\zeta+c_4\xi\eta+c_5\xi\zeta+c_6\eta\zeta+c_7\xi\eta\zeta
\end{aligned} \tag{3-97}$$

Using strain-displacement relation, the strains are derived as follows,

$$
\begin{aligned}
\varepsilon_{\xi}&=a_1+a_4\eta+a_5\zeta+a_7\eta\zeta\\
\varepsilon_{\eta}&=b_2+b_4\xi+b_6\zeta+b_7\xi\zeta\\
\varepsilon_{\zeta}&=c_3+c_5\xi+c_6\eta+c_7\xi\eta\\
\varepsilon_{\xi\eta}&=(a_2+b_1)+a_4\xi+b_4\eta+(a_6+b_5)\zeta+a_7\xi\zeta+b_7\eta\zeta\\
\varepsilon_{\eta\zeta}&=(c_2+b_3)+(c_4+b_5)\xi+b_6\eta+c_6\zeta+b_7\xi\eta+c_7\xi\zeta\\
\varepsilon_{\zeta\xi}&=(c_1+a_3)+a_5\xi+(c_4+a_6)\eta+c_5\zeta+a_7\xi\eta+c_7\eta\zeta
\end{aligned}
\tag{3-98}
$$

For linear elasticity body, the material stiffness matrix is

$$
[C]=\begin{bmatrix}
C_{11} & C_{12} & C_{13} & C_{14} & C_{15} & C_{16}\\
C_{21} & C_{22} & C_{23} & C_{24} & C_{25} & C_{26}\\
C_{31} & C_{32} & C_{33} & C_{34} & C_{35} & C_{36}\\
C_{41} & C_{42} & C_{43} & C_{44} & C_{45} & C_{46}\\
C_{51} & C_{52} & C_{53} & C_{54} & C_{55} & C_{56}\\
C_{61} & C_{62} & C_{63} & C_{64} & C_{65} & C_{66}
\end{bmatrix}
\tag{3-99}
$$

Using the constitutive relation (3-76), the displacement-derived stress field can be obtained,

$$
\bar{\boldsymbol{\sigma}}=[\boldsymbol{\Theta}]\boldsymbol{\gamma}=\begin{bmatrix}
[\boldsymbol{\Theta}_1] & 0 & 0 & 0 & 0 & 0\\
0 & [\boldsymbol{\Theta}_1] & 0 & 0 & 0 & 0\\
0 & 0 & [\boldsymbol{\Theta}_1] & 0 & 0 & 0\\
0 & 0 & 0 & [\boldsymbol{\Theta}_1] & 0 & 0\\
0 & 0 & 0 & 0 & [\boldsymbol{\Theta}_1] & 0\\
0 & 0 & 0 & 0 & 0 & [\boldsymbol{\Theta}_1]
\end{bmatrix}\boldsymbol{\gamma}
\tag{3-100}
$$

In which,

$$
[\boldsymbol{\Theta}_1]=[\;1\;\;\xi\;\;\eta\;\;\zeta\;\;\xi\eta\;\;\eta\zeta\;\;\zeta\xi\;]
\tag{3-101}
$$

Using iso-function relation (3-80), an assumed stress field is obtained,

$$\boldsymbol{\sigma} = [P]\,\boldsymbol{\beta} = [\Theta]\,\boldsymbol{\beta} = \begin{bmatrix} [\Theta_1] & 0 & 0 & 0 & 0 & 0 \\ 0 & [\Theta_1] & 0 & 0 & 0 & 0 \\ 0 & 0 & [\Theta_1] & 0 & 0 & 0 \\ 0 & 0 & 0 & [\Theta_1] & 0 & 0 \\ 0 & 0 & 0 & 0 & [\Theta_1] & 0 \\ 0 & 0 & 0 & 0 & 0 & [\Theta_1] \end{bmatrix} \boldsymbol{\beta} \tag{3-102}$$

Within the matrix $[\Theta]$ (3-101), there are seven least-order polynomial terms. Therefore, the assumed stress field (3-102) contains forty two stress parameters. These stress parameters are basic variables in the variational functional and are determined by the variational principle.

For partial hybrid element, the displacement-derived partial stress field is

$$\overline{\boldsymbol{\sigma}_g} = [\Theta_g]\,\boldsymbol{\gamma} = \begin{bmatrix} [\Theta_1] & 0 & 0 \\ 0 & [\Theta_1] & 0 \\ 0 & 0 & [\Theta_1] \end{bmatrix} \boldsymbol{\gamma} \tag{3-103}$$

Using iso-function relation (3--83), an assumed partial stress field can be obtained as follows,

$$\boldsymbol{\sigma}_g = [P_g]\,\boldsymbol{\beta} = [\Theta_g]\,\boldsymbol{\beta} = \begin{bmatrix} [\Theta_1] & 0 & 0 \\ 0 & [\Theta_1] & 0 \\ 0 & 0 & [\Theta_1] \end{bmatrix} \boldsymbol{\beta} \tag{3-104}$$

In this partial stress field, there are twenty one stress parameters only. Comparing to the assumed stress field (3-102), the number of stress parameters is greatly reduced. For analysis of composite structures, it will greatly improve the efficiency of finite element models of the structures.

The eigenvalue examination shows that conventional hybrid element and partial hybrid element have the same eigenvalue property as their displacement counterpart. This result can be expected according to the limitation principle [3.30-3.32] for hybrid elements and partial hybrid elements presented above.

3.4.3 Classification Method

Although the iso-function method [3.6-3.7] is used easily for establishing assumed stress fields, there are a great number of unnecessary stress parameters and stress modes in the stress fields. Due to the inversion of the flexibility matrix [H], unnecessary stress modes will require more and unnecessary computer resources and the efficiency of finite element model is reduced. On the other hand, eigenfunction method [3.5, 3.29] can establish an assumed stress field which only contains minimum number of stress modes. But, using this method, multiple eigenvalues will cause difficulty. In the eigenfunction method, theoretically, there should be a unique set of natural stress modes in a hybrid element. In practice, it is difficult to find them due to existence of multiple eigenvalues. When multiple eigenvalues exist, there will be many choices for the directions of the corresponding eigenvectors and the resulting stress modes by the eigenvectors are not unique. Therefore, the assumption in eigenfunction method has to be modified. In this section, a classification method [3.16] is presented to establish assumed stress fields based on the iso-function method and the eigenfunction method.

Classification of Stress Modes

Since Pian [3.33] formulated a hybrid element in 1964, a lot of different hybrid elements have been presented. However, a hybrid element may have many different assumed stress fields. For example, there are many assumed stress fields for 2-D, 4-node plane element and 3-D, 8-node solid element. Pian [3.22] proposed an assumed stress field for 2-D, 4-node plane element and another for 3-D, 8-node solid element. Punch and Atluri [3.34-3.35] gave two assumed stress fields for 2-D, 4-node plane element, and eight assumed stress fields for 3-D, 8-node solid element. Huang [3.5] presented an assumed stress field for 3-D, 8-node solid element. Although each of these assumed stress fields may contain the same number of stress modes, the stress modes in these fields are different. In order to determine the optimal stress modes for an assumed stress field, it is necessary to study the relationship between different stress modes.

In section 3.4.1, it is assumed that a finite element has (n-r) natural deformation modes and r rigid body modes, and the displacement distribution in the element can be represented by them. It is also assumed that there exists a unique set of natural stress modes in an element, and they can be determined by (n-r) natural deformation modes. Therefore, the eigenfunction method is used to search these natural stress modes. Although the eigenfunction method may fail to find the natural stress modes in the case of multiple eigenvalues, it can be modified to classify stress modes that appeared in various assumed stress fields.

In a hybrid element, if various stress modes can be classified, at least m stress mode groups must exist because the stiffness matrix of hybrid element must have m non-zero eigenvalues, except zero-energy stress mode group. Otherwise, the hybrid element will contain kinematic deformation modes. On the other hand, no matter how many stress modes there are in a stress matrix [P], the maximum number of non-zero eigenvalues of an element stiffness matrix is always equal to or less than m. Therefore, the number of stress mode groups is equal to or less than m. Thus, it can be considered that there exist and only exist m stress mode groups except zero-energy modes for a hybrid element.

In addition, the eigenvectors and eigenvalues of the stiffness matrix will be sensitive to the assumed stress modes. The eigenvalue examination will give r zero eigenvalues corresponding to rigid body modes and m (= n-r) non-zero eigenvalues corresponding to natural deformation modes if the assumed stress field is suitable. Therefore, m stress mode groups must be related to m natural deformation modes and the zero-energy stress mode group must be corresponding to rigid body modes. Thus, all stress modes in various assumed stress matrices can be classified into the m+1 stress mode groups.

Postulate 2

There exist and only exist m (=n-r) natural deformation modes in a hybrid element. All stress modes in assumed stress field can be classified into m stress mode groups corresponding to m natural deformation modes and a zero energy mode group corresponding to rigid body modes of the element which has n degrees of freedom and r rigid body modes.

Based on this postulate, it can be considered that an assumed stress field can be represented by stress modes in the m stress mode groups related to m natural deformation modes. This can be expressed as follows,

$$\{\sigma\}=[P]\{\beta\}=[\{\sigma_1\},\{\sigma_2\}, \;...\; ,\{\sigma_m\}]\begin{Bmatrix}\beta_1\\ \beta_2\\ \cdot\\ \cdot\\ \beta_m\end{Bmatrix}=\sum_{i=1}^{m}[P_i]\{\beta_i\} \tag{3-105}$$

where $[P_i]$ and $\{\beta_i\}$ (i=1,2,...m) are the stress matrices and stress-coefficient vectors related to the i-th stress mode group which corresponds to the i-th natural deformation mode. They are

$$[P_i]=[\{0\}\;...\;\{0\},\{\sigma_i\},\{0\}\;...\;\{0\}] \tag{3-106}$$

and

$$\{\boldsymbol{\beta}_i\} = [\ 0\ \ldots\ 0\ \boldsymbol{\beta}_i\ 0\ \ldots\ 0\ ]^T \tag{3-107}$$

The stress mode which belongs to the i-th stress mode group can be expressed in the form,

$$\{\overline{\sigma_i}\} = [P]\{\boldsymbol{\beta}_i\} \tag{3-108}$$

Therefore, the vector $\{\beta_i\}$ corresponds the i-th stress mode group which corresponds to the natural deformation mode $\{\delta_i\}$ (i=1,2,...m). Using equation (2-31), we have

$$[H]\ \{\boldsymbol{\beta}_i\} = [G]\ \{\boldsymbol{\delta}_i\} \tag{3-109}$$

If the stress matrix [P] does not contain any stress mode which belongs to the i-th stress mode group, the value of β_i in the vector $\{\beta_i\}$ should be equal to zero. Then, one can add a new stress mode into the stress matrix [P]. The new stress mode will be classified by m natural deformation modes. Corresponding to the i-th natural deformation mode $\{\delta_i\}$, the condition to check whether the new stress mode belongs to the i-th stress mode group can be expressed in the form,

$\beta_i = 0$ if new stress mode does not belong to i-th stress mode group

$\beta_i \neq 0$ if new stress mode belongs to i-th stress mode group

Using equations (3-39), (3-40), and (3-109), the eigenvalues are obtained as follows,

$$\boldsymbol{\lambda}_i = \{\boldsymbol{\delta}_i\}^T [K] \{\boldsymbol{\delta}_i\} = \{\boldsymbol{\beta}_i\}^T [H] \{\boldsymbol{\beta}_i\} \tag{3-110}$$

Because all of the diagonal elements in the flexibility matrix [H] may not be equal to zero, the classification condition above becomes

$\lambda_i = 0$ if new stress mode does not belong to i-th stress mode group

$\lambda_i \neq 0$ if new stress mode belongs to i-th stress mode group

Thus, using eigenvalue examination, the stress modes can be classified into m+1 stress mode groups.

Expression of Classification Condition of Stress Modes

A hybrid element stiffness matrix [K] can be formulated using equations (2-28) and (2-34). Its eigenvalues and eigenvectors are calculated from equation (3-42). The eigenvectors $\{\delta_i\}$ (i=1,2,...m) satisfy the condition (3-40). Thus, the eigenvalue equation (3-42) is changed to

$$\lambda_i=\{\delta_i\}^T[K]\{\delta_i\} \tag{3-111}$$

For any stress mode $\{\sigma_j\}$ among m stress modes $\{\sigma_1, \sigma_2, ..., \sigma_m\}$, the stiffness matrix $[K_j]$ can be derived using equations (3-43) and (3-44). Corresponding to the i-th natural deformation mode, one has

$$\lambda_i=\{\delta_i\}^T[K_j]\{\delta_i\} \tag{3-112}$$

According to the classification condition of stress modes, if the stress mode $\{\sigma_j\}$ belongs to the i-th stress mode group corresponding to the natural deformation mode $\{\delta_i\}$, the eigenvalue λ_i is a non-zero value; otherwise, the eigenvalue λ_i equals zero. This condition can be expressed in the form,

$$\begin{aligned} &\lambda_i=\{\delta_i\}^T[K_j]\{\delta_i\}=0 \qquad && i\neq j \\ &\lambda_i=\{\delta_i\}^T[K_j]\{\delta_i\}\neq 0 \qquad && i=j \end{aligned} \tag{3-113}$$

If the stress mode $\{\sigma_j\}$ is a zero-energy stress mode, all eigenvalues λ_i (i=1,2,...,m) equal zero.

In the section 3.4.1, there is a postulate 1. This postulate can be stated as a theorem as follows.

Theorem 2

If and only if the flexibility matrix [H] is a diagonal matrix, the eigenvalues obtained from the separate stress mode equations

$$([K_i]-\lambda[I])\{\delta\}=0, \qquad i=1,2,...,m \tag{3-114}$$

are equal to the eigenvalues obtained from the total stress mode equation

$$([K]-\lambda[I])\{\delta\}=0 \tag{3-115}$$

in which, the matrices $[K_i]$ and [K] are defined in equations (2-28), (2-34),

(3-43), and (3-44).

Proof:

From the equations (3-111), (2-28) and (3-109), one has

$$\lambda=\{\delta_i\}^T[K]\{\delta_i\}=\{\beta_i\}^T[H]\{\beta_i\} \tag{3-116}$$

Because the matrix [H] is a diagonal matrix, one has

$$[H]=\sum_{j=1}^{m}[H_j] \tag{3-117}$$

Thus, using equations (3-106), (3-107) and (3-43), the eigenvalue of the matrix [K] becomes

$$\lambda=\sum_{j=1}^{m}\{\beta_i\}^T[H_j]\{\beta_i\}=\{\beta_i\}^T[H_i]\{\beta_i\} \tag{3-118}$$

Furthermore, using equations (3-109) and (3-113), this equation becomes

$$\lambda=\{\beta_i\}^T[H_i]\{\beta_i\}=\{\delta_i\}^T[K_i]\{\delta_i\}=\lambda_i \tag{3-119}$$

End of proof.

By means of the theorem 1, it has ben proven that if the flexibility matrix [H] is a diagonal matrix, the energy of the element is decomposable. Therefore, the theorem 2 is equivalent to the postulate 1.

<u>*Theorem 3*</u>

If and only if the flexibility matrix [H] is a diagonal matrix, the classification of stress modes is unique.

Proof:

If a stress mode among m stress modes that form the stress matrix [P] appears in more than one stress mode group, one of the m stress mode groups must contain two stress modes. Assume that a stress mode $\{\sigma_j\}$ in the j-th stress mode group also appears in the i-th stress mode group. Thus, at least two stress modes $\{\sigma_i\}$ and $\{\sigma_j\}$ will belong to the i-th stress mode group corresponding to the natural deformation modes $\{\delta_i\}$. Therefore, one has

$$\lambda_{ii}=\{\delta_i\}^T\ [K_i]\{\delta_i\} \quad and \quad \lambda_{ij}=\{\delta_i\}^T\ [K_j]\{\delta_i\} \tag{3-120}$$

Corresponding to the natural deformation mode $\{\delta_i\}$, one can obtain the eigenvalue of the stiffness matrix [K] formulated by m stress modes as follows,

$$\lambda=\{\delta_i\}^T\ [K]\{\delta_i\} \tag{3-121}$$

Because the flexibility matrix [H] is diagonal, the energy of the element is decomposable and the stiffness matrix satisfies the superposition principle. From equations (3-45) and (3-120), we obtain

$$\lambda=\{\delta_i\}^T\ [K]\{\delta_i\}=\sum_{k=1}^{m}\{\delta_i\}^T\ [K_k]\{\delta_i\}=\lambda_{ii}+\lambda_{ij} \tag{3-122}$$

using theorem 2, one has

$$\lambda = \lambda_{ii} \tag{3-123}$$

From the equation (3-122) and (3-123), we obtain

$$\lambda_{ij} = 0 \tag{3-124}$$

According to the condition of classification, the stress modes $\{\sigma_j\}$ does not belong to the i-th stress mode group. Therefore, the i-th stress mode group only contains $\{\sigma_i\}$ and the stress modes $\{\sigma_j\}$ can not appear in two stress mode groups. Thus, if the matrix [H] is diagonal, the classification of m stress modes is unique.

End of proof

Determination of Optimal Stress Matrix

Before classifying stress modes, one can find a number of initial stress modes since there are many approaches to derive assumed stress matrices for a hybrid element. For example, Pian and Chen [3.22] used the product $\{\sigma\}^T\{\varepsilon\}$ to determine the necessary assumed stress modes. Punch and Atluri [3.34-3.35] used group theory to obtain assumed stress matrices. One can also derive an assumed stress matrix using iso-function method.

In order to present a systematic procedure for classifying stress modes and constructing assumed stress fields, the iso-function method is used to derive initial stress modes to be classified in this work. This is because the hybrid element constructed by the iso-function stress matrix has the same eigenvalues and eigenvectors as its conventional displacement counterpart. Also, the method using iso-function is straightforward and can be followed easily. After obtaining initial stress modes, one can use eigenvalue examination to find m representative stress modes that represent m stress mode groups corresponding to m natural deformation modes. The stress matrix consisted of the m representative stress modes is an optimal stress matrix. Then, all existing stress modes can be classified into m+1 stress mode groups. Its detail is presented as follows,

Step 1:

Derive an initial stress matrix $[P]_{iso}$ by iso-function method. The number of initial stress modes in the matrix is always larger than m (=n-r). In order to select m necessary stress modes, these initial stress modes have to be classified into (n-r) stress mode groups.

Step 2:

Select stress modes in the order from low order term to high order term. Now select a stress mode from the existing stress matrix $[P]_{iso}$ and form an assumed stress matrix $[P_1]$. The element stiffness matrix [K] corresponding to stress matrix $[P_1]$ can be obtained by using equations (2-28) and (2-34). If the eigenvalue examination gives a non-zero eigenvalue, the stress mode is a non-zero-energy stress mode; otherwise, it is a zero-energy stress mode. Repeating the eigenvalue examination to check whether a stress mode is a zero-energy stress mode for all stress modes in the existing stress matrix $[P]_{iso}$.

Take all zero-energy stress modes out and keep non-zero-energy stress modes in the matrix $[P]_{iso}$. All zero-energy stress modes form a zero-energy stress mode group.

Step 3:

Take a non-zero-energy stress mode from the existing stress matrix $[P]_{iso}$ and form an assumed stress matrix $[P_1]$. The stress mode $\{\sigma_1\}$ is the representative stress mode which represents group 1 of stress modes.

Step 4:

Add another stress mode selected in the existing stress matrix $[P]_{iso}$ into the assumed stress matrix $[P_1]$ and form a new stress matrix $[P_2]$,

$$[P_2] = [\{\sigma_1\}\ \{\sigma_2\}] \tag{3-125}$$

Step 5:

The eigenvalue examination gives the eigenvalues of the stiffness matrix. If there is only one non-zero eigenvalue, continue to step 6. If there are two non-zero eigenvalues, go to step 7.

Step 6:

In this case, the added stress mode belongs to group 1 of stress modes. Take the second stress mode out and put it in group 1 of stress modes. Then, go back to step 4.

Step 7:

The two stress modes belong to two different groups of stress modes. The second stress mode $\{\sigma_2\}$ is the representative stress mode which represents group 2 of stress modes.

Step 8:

Add another stress mode selected from the matrix $[P]_{iso}$ into the assumed stress matrix $[P_2]$ and form a new stress matrix $[P_3]$,

$$[P_3] = [\{\sigma_1\}\ \{\sigma_2\}\ \{\sigma_3\}] \tag{3-126}$$

Step 9:

The element stiffness matrix [K] and its eigenvalues are calculated. If there are only two non-zero eigenvalues, continue to step 10. If there are three non-zero eigenvalues, go to step 11.

Step 10:

In this case, the new stress mode $\{\sigma_3\}$ belongs to one of the two stress mode groups. Construct the matrices $[P'_2]$ and $[P''_2]$ as follows,

$$[P_2'] = [\{\sigma_1\}\ \{\sigma_3\}] \quad or \quad [P_2''] = [\{\sigma_2\}\ \{\sigma_3\}] \tag{3-127}$$

If the stiffness matrix corresponding to the stress matrix $[P_2']$ has two non-zero eigenvalues, the stress mode $\{\sigma_3\}$ belongs to the group 2 of stress modes. Otherwise, the stress mode $\{\sigma_3\}$ belongs to the group 1 of stress modes. Put the stress mode $\{\sigma_3\}$ into the corresponding stress mode group, and go back to step 8.

Step 11:

In this case, the three stress modes belong to three different stress mode groups. The added stress mode $\{\sigma_3\}$ is the representative stress mode which represents group 3 of stress modes.

Step 12:

Add one more stress mode selected from the matrix $[P]_{iso}$ into the matrix $[P_3]$ and form a new stress matrix $[P_4]$,

$$[P_4] = [\{\sigma_1\}\ \{\sigma_2\}\ \{\sigma_3\}\ \{\sigma_4\}] \tag{3-128}$$

and so on. Repeating the same process until m representative stress modes that represent m stress mode groups are obtained. The m(=n-r) representative stress modes correspond to m natural deformation modes and form a optimal stress matrix $[P]_{opt}$ from the existing stress matrix $[P]_{iso}$.

Classification of Other Stress Modes

Step 13:

After m representative stress modes are obtained, other initial stress modes that remain in the existing stress matrix $[P]_{iso}$ can be classified into the m stress mode groups. Many other stress modes derived by different methods also can be classified into the m stress mode groups corresponding to m natural deformation modes and the zero-energy stress mode group corresponding to rigid body modes.

Based on the optimal stress matrix $[P]_{opt}$, any remaining stress mode in $[P]_{iso}$ can be classified by using it to replace each and every stress mode in the matrix $[P]_{opt}$ in order. Once the eigenvalue examination results in m non-zero eigenvalues, the representative stress mode which is replaced and the one which replaces it belong to the same stress mode group. Put the remaining stress mode into the corresponding stress mode group and recover the optimal stress matrix $[P]_{opt}$. Then, classify another remaining stress mode.

Step 14:

Repeating the same process until all remaining stress modes are classified. Thus, all existing stress modes are classified into m+1 different mode groups. Every stress mode group contains many interchangeable stress modes. For a stress mode derived by other method, if eigenvalue examination always give m-1 non-zero eigenvalues when this stress mode replaces each and every stress mode in the matrix $[P]_{opt}$, this stress mode is a zero-energy stress mode.

Illustration for the Classification of Stress Modes

As an illustration for the above procedure, the stress modes presented in ref. [3.5,3.22,3.34-3.35] and those derived by iso-function method are classified.

2-D, 4-node Plane Hybrid Element

1. Determination of optimal stress matrix

The 2-D, 4-node plane element has (n=) 8 degrees of freedom and (r=) 3 rigid body modes. So it has (m=n-r=) 5 natural deformation modes. Firstly, an assumed stress matrix can be derived from the assumed displacement field of the element by the iso-function method,

$$[P_I] = \begin{bmatrix} 1 & 0 & 0 & x & y & 0 & 0 & 0 & 0 \\ 0 & 1 & 0 & 0 & 0 & x & y & 0 & 0 \\ 0 & 0 & 1 & 0 & 0 & 0 & 0 & x & y \end{bmatrix} \tag{3-129}$$

The number of stress modes in the stress matrix is larger than m (=5). The stress matrix derived by iso-function method contains a few unnecessary stress modes. The eigenvalue examination indicates that the eigenvalues and eigenvectors of the hybrid element stiffness matrix constructed by the assumed stress matrix $[P_I]$ are the same as that of displacement element stiffness matrix. Therefore, the stress modes in the stress matrix are taken as initial stress modes to be classified. There are nine stress modes in the matrix $[P_I]$,

$$\{\sigma_1\}=\begin{Bmatrix}1\\0\\0\end{Bmatrix} \quad \{\sigma_2\}=\begin{Bmatrix}0\\1\\0\end{Bmatrix} \quad \{\sigma_3\}=\begin{Bmatrix}0\\0\\1\end{Bmatrix}$$

$$\{\sigma_4\}=\begin{Bmatrix}x\\0\\0\end{Bmatrix} \quad \{\sigma_5\}=\begin{Bmatrix}y\\0\\0\end{Bmatrix} \quad \{\sigma_6\}=\begin{Bmatrix}0\\x\\0\end{Bmatrix} \qquad (3\text{-}130)$$

$$\{\sigma_7\}=\begin{Bmatrix}0\\y\\0\end{Bmatrix} \quad \{\sigma_8\}=\begin{Bmatrix}0\\0\\x\end{Bmatrix} \quad \{\sigma_9\}=\begin{Bmatrix}0\\0\\y\end{Bmatrix}$$

It will save computation time for calculating element stiffness matrix if the number of the stress modes can be reduced to m (=n-r). In order to do it, the initial stress modes in the existing stress matrix have to be classified into m stress mode groups. First of all, one must find m representative stress modes corresponding to m natural deformation modes. Following step 2 - step 12 in the procedure of the classification method given in the above section, one can obtain 5 representative stress modes $\{\sigma_1\ \sigma_2\ \sigma_3\ \sigma_5\ \sigma_6\}$ corresponding to (m=) 5 natural deformation modes and the zero-energy stress modes $\{\sigma_4\}$ and $\{\sigma_7\}$ corresponding to rigid body modes. The eigenvalues of the stiffness matrix related to $\{\sigma_1\ \sigma_2\ \sigma_3\ \sigma_5\ \sigma_6\}$ are not equal to zero, and the eigenvalue of stiffness matrix related to $\{\sigma_4\}$ or $\{\sigma_7\}$ is equal to zero. The 5 representative stress modes form an optimal stress matrix $[P_{II}]$ from the existing stress matrix $[P_I]$,

$$[P_{II}] = [\sigma_1\ \ \sigma_2\ \ \sigma_3\ \ \sigma_5\ \ \sigma_6] = \begin{bmatrix}1 & 0 & 0 & y & 0\\ 0 & 1 & 0 & 0 & x\\ 0 & 0 & 1 & 0 & 0\end{bmatrix} \qquad (3\text{-}131)$$

The stress matrix is the same as that given by Pian [3.22].

2. Classification of other stress modes

After obtaining the optimal stress matrix, one can classify stress modes in the existing stress matrix $[P_I]$ into (m+1=) 6 stress mode groups by following step 13 to step 14 in the procedure,

Tension mode (Group 1): $\{\ \sigma_1\ \}$

Tension mode (Group 2): $\{\sigma_2\}$

Shear mode (Group 3): $\{\sigma_3\}$

Bending mode (Group 4): $\{\sigma_5\}$, $\{\sigma_8\}$

Bending mode (Group 5): $\{\sigma_6\}$, $\{\sigma_9\}$

Zero-energy stress mode (Group 6): $\{\sigma_4\}$, $\{\sigma_7\}$

The first 5 stress mode groups correspond to 5 natural deformation modes and the zero- energy stress mode group corresponds to rigid body modes.

There are many methods to derive initial stress modes. For example, in the two assumed stress matrices derived by means of group theory[3.34,3.35] for the same finite element, there are 4 stress modes that are different from stress modes $\{\sigma_1\}$-$\{\sigma_9\}$above:

$$\{\sigma_{10}\}=\begin{Bmatrix}1\\1\\0\end{Bmatrix} \quad \{\sigma_{11}\}=\begin{Bmatrix}1\\-1\\0\end{Bmatrix} \quad \{\sigma_{12}\}=\begin{Bmatrix}0\\-y\\x\end{Bmatrix} \quad \{\sigma_{13}\}=\begin{Bmatrix}-x\\0\\y\end{Bmatrix} \tag{3-132}$$

Moreover, one may want to introduce some stress modes of high order term into the assumed stress matrix [P] in order to describe special stress distribution in a local region of a structure to be solved. For example,

$$\begin{gathered}
\{\sigma_{14}\}=\begin{Bmatrix}x^2\\0\\0\end{Bmatrix} \quad \{\sigma_{15}\}=\begin{Bmatrix}0\\x^2\\0\end{Bmatrix} \quad \{\sigma_{16}\}=\begin{Bmatrix}0\\0\\x^2\end{Bmatrix} \\
\{\sigma_{17}\}=\begin{Bmatrix}y^2\\0\\0\end{Bmatrix} \quad \{\sigma_{18}\}=\begin{Bmatrix}0\\y^2\\0\end{Bmatrix} \quad \{\sigma_{19}\}=\begin{Bmatrix}0\\0\\y^2\end{Bmatrix} \\
\{\sigma_{20}\}=\begin{Bmatrix}xy\\0\\0\end{Bmatrix} \quad \{\sigma_{21}\}=\begin{Bmatrix}0\\xy\\0\end{Bmatrix} \quad \{\sigma_{22}\}=\begin{Bmatrix}0\\0\\xy\end{Bmatrix}
\end{gathered} \tag{3-133}$$

According to the steps 13 - 14 in the procedure of classification method, these new stress modes $\{\sigma_{10}\}$-$\{\sigma_{22}\}$ can also be classified into the 6 stress mode groups above,

Tension mode (Group 1): $\{ \sigma_1 \}, \{ \sigma_{10} \}, \{ \sigma_{14} \}, \{ \sigma_{17} \}$

Tension mode (Group 2): $\{ \sigma_2 \}, \{ \sigma_{11} \}, \{ \sigma_{15} \}, \{ \sigma_{18} \}$

Shear mode (Group 3): $\{ \sigma_3 \}, \{ \sigma_{16} \}, \{ \sigma_{19} \}$

Bending mode (Group 4): $\{ \sigma_5 \}, \{ \sigma_8 \}, \{ \sigma_{12} \}$

Bending mode (Group 5): $\{ \sigma_6 \}, \{ \sigma_9 \}, \{ \sigma_{13} \}$

Zero-energy stress mode (Group 6): $\{ \sigma_4 \}, \{ \sigma_7 \}, \{ \sigma_{20} \}, \{ \sigma_{21} \}, \{ \sigma_{22} \}$

More high-order stress modes can be classified into the 6 stress mode groups above by using the classification method. If the flexibility matrix [H] is a diagonal matrix, the classification of the stress modes is unique.

3-D, 8-node Solid Hybrid Element

1. Determination of optimal stress matrix

The 3-D, 8-node solid element has (n=) 24 degrees of freedom and (r=) 6 rigid body modes. So it has (m=n-r=) 18 natural deformation modes. By means of iso-function method, an initial stress matrix [P] can be derived from the assumed displacement field of the element as follows

$$[P]_{ISO} = \left| \begin{array}{cccccccccccccccccccccccc}
1&0&0&0&0&0&x&0&0&0&0&0&y&0&0&0&0&0&z&0&0&0&0&0\\
0&1&0&0&0&0&0&x&0&0&0&0&0&y&0&0&0&0&0&z&0&0&0&0\\
0&0&1&0&0&0&0&0&x&0&0&0&0&0&y&0&0&0&0&0&z&0&0&0\\
0&0&0&1&0&0&0&0&0&x&0&0&0&0&0&y&0&0&0&0&0&z&0&0\\
0&0&0&0&1&0&0&0&0&0&x&0&0&0&0&0&y&0&0&0&0&0&z&0\\
0&0&0&0&0&1&0&0&0&0&0&x&0&0&0&0&0&y&0&0&0&0&0&z
\end{array} \right.$$

$$\left. \begin{array}{ccccccccccccccc}
xy&0&0&0&0&yz&0&0&0&0&zx&0&0&0&0\\
0&xy&0&0&0&0&yz&0&0&0&0&zx&0&0&0\\
0&0&xy&0&0&0&0&yz&0&0&0&0&zx&0&0\\
0&0&0&0&0&0&0&0&yz&0&0&0&0&zx&0\\
0&0&0&xy&0&0&0&0&0&0&0&0&0&0&zx\\
0&0&0&0&xy&0&0&0&0&yz&0&0&0&0&0
\end{array} \right| \qquad (3\text{-}134)$$

There are 39 stress modes to be classified in the stress matrix. The number of stress modes is larger than m (=18). The eigenvalue examination shows that the eigenvalues and eigenvectors of the hybrid element stiffness matrix formulated by the assumed stress matrix $[P]_{ISO}$ are the same as that of displacement element stiffness matrix. The stress modes in the matrix $[P]_{ISO}$ are taken as initial stress modes to be classified. The 39 stress modes in the matrix $[P]_{ISO}$ are numbered as follows:

$$\{\sigma_1\ \sigma_2\ \sigma_3\ \sigma_4\ \sigma_5\ \sigma_6\} = \begin{Bmatrix} 1 & 0 & 0 & 0 & 0 & 0 \\ 0 & 1 & 0 & 0 & 0 & 0 \\ 0 & 0 & 1 & 0 & 0 & 0 \\ 0 & 0 & 0 & 1 & 0 & 0 \\ 0 & 0 & 0 & 0 & 1 & 0 \\ 0 & 0 & 0 & 0 & 0 & 1 \end{Bmatrix} \tag{3-135}$$

$$\{\sigma_7\ \sigma_8\ \sigma_9\ \sigma_{10}\ \sigma_{11}\ \sigma_{12}\} = \begin{Bmatrix} x & 0 & 0 & 0 & 0 & 0 \\ 0 & x & 0 & 0 & 0 & 0 \\ 0 & 0 & x & 0 & 0 & 0 \\ 0 & 0 & 0 & x & 0 & 0 \\ 0 & 0 & 0 & 0 & x & 0 \\ 0 & 0 & 0 & 0 & 0 & x \end{Bmatrix} \tag{1-136}$$

$$\{\sigma_{13}\ \sigma_{14}\ \sigma_{15}\ \sigma_{16}\ \sigma_{17}\ \sigma_{18}\} = \begin{Bmatrix} y & 0 & 0 & 0 & 0 & 0 \\ 0 & y & 0 & 0 & 0 & 0 \\ 0 & 0 & y & 0 & 0 & 0 \\ 0 & 0 & 0 & y & 0 & 0 \\ 0 & 0 & 0 & 0 & y & 0 \\ 0 & 0 & 0 & 0 & 0 & y \end{Bmatrix} \tag{1-137}$$

$$\{\sigma_{19}\ \sigma_{20}\ \sigma_{21}\ \sigma_{22}\ \sigma_{23}\ \sigma_{24}\} = \left\{\begin{matrix} z & 0 & 0 & 0 & 0 & 0 \\ 0 & z & 0 & 0 & 0 & 0 \\ 0 & 0 & z & 0 & 0 & 0 \\ 0 & 0 & 0 & z & 0 & 0 \\ 0 & 0 & 0 & 0 & z & 0 \\ 0 & 0 & 0 & 0 & 0 & z \end{matrix}\right\} \tag{1-138}$$

$$\{\sigma_{25}\ \sigma_{26}\ \sigma_{27}\ \sigma_{28}\ \sigma_{29}\} = \left\{\begin{matrix} xy & 0 & 0 & 0 & 0 \\ 0 & xy & 0 & 0 & 0 \\ 0 & 0 & xy & 0 & 0 \\ 0 & 0 & 0 & 0 & 0 \\ 0 & 0 & 0 & xy & 0 \\ 0 & 0 & 0 & 0 & xy \end{matrix}\right\} \tag{1-139}$$

$$\{\sigma_{30}\ \sigma_{31}\ \sigma_{32}\ \sigma_{33}\ \sigma_{34}\} = \left\{\begin{matrix} yz & 0 & 0 & 0 & 0 \\ 0 & yz & 0 & 0 & 0 \\ 0 & 0 & yz & 0 & 0 \\ 0 & 0 & 0 & yz & 0 \\ 0 & 0 & 0 & 0 & 0 \\ 0 & 0 & 0 & 0 & yz \end{matrix}\right\} \tag{1-140}$$

$$\{\sigma_{35}\ \sigma_{36}\ \sigma_{37}\ \sigma_{38}\ \sigma_{39}\} = \left\{\begin{matrix} zx & 0 & 0 & 0 & 0 \\ 0 & zx & 0 & 0 & 0 \\ 0 & 0 & zx & 0 & 0 \\ 0 & 0 & 0 & zx & 0 \\ 0 & 0 & 0 & 0 & zx \\ 0 & 0 & 0 & 0 & 0 \end{matrix}\right\} \tag{1-141}$$

These stress modes are classified one by one in the order from low order term to high order term. Following step 2 - step 12 in the procedure of the classification, one can obtain (m=) 18 representative stress modes { σ_1 σ_2 σ_3 σ_4 σ_5 σ_6 σ_8 σ_9 σ_{11} σ_{13} σ_{15} σ_{18} σ_{19} σ_{20} σ_{22} σ_{27} σ_{30} σ_{36} } corresponding to 18 natural deformation modes. These representative stress modes form an optimal stress matrix $[P_1]$ from the existing stress matrix $[P]_{iso}$ as follows:

$$[P_1] = [\sigma_1 \ \sigma_2 \ \sigma_3 \ \sigma_4 \ \sigma_5 \ \sigma_6 \ \sigma_{13} \ \sigma_{20} \ \sigma_9 \ \sigma_{19} \ \sigma_8 \ \sigma_{15} \ \sigma_{22} \ \sigma_{11} \ \sigma_{18} \ \sigma_{30} \ \sigma_{36} \ \sigma_{27}] \tag{1-142}$$

$$[P_1] = \begin{bmatrix} 1 & 0 & 0 & 0 & 0 & 0 & y & 0 & 0 & z & 0 & 0 & 0 & 0 & 0 & yz & 0 & 0 \\ 0 & 1 & 0 & 0 & 0 & 0 & 0 & z & 0 & 0 & x & 0 & 0 & 0 & 0 & 0 & zx & 0 \\ 0 & 0 & 1 & 0 & 0 & 0 & 0 & 0 & x & 0 & 0 & y & 0 & 0 & 0 & 0 & 0 & xy \\ 0 & 0 & 0 & 1 & 0 & 0 & 0 & 0 & 0 & 0 & 0 & 0 & z & 0 & 0 & 0 & 0 & 0 \\ 0 & 0 & 0 & 0 & 1 & 0 & 0 & 0 & 0 & 0 & 0 & 0 & 0 & x & 0 & 0 & 0 & 0 \\ 0 & 0 & 0 & 0 & 0 & 1 & 0 & 0 & 0 & 0 & 0 & 0 & 0 & 0 & y & 0 & 0 & 0 \end{bmatrix} \tag{1-143}$$

This stress matrix is the same as that proposed by Pian [3.22].

2. Classification of other stress modes

Following steps 13 - step 14 in the procedure, other stress modes that remain in the existing stress matrix $[P]_{iso}$ can be classified into m+1 (=19) stress mode groups as follows:

Tension and compression modes (3 groups): $[\{\sigma_1\}_{G1}, \{\sigma_2\}_{G2}, \{\sigma_3\}_{G3}]$

Pure shear modes (3 groups): $[\{\sigma_4\}_{G4}, \{\sigma_5\}_{G5}, \{\sigma_6\}_{G6}]$

Bending modes (6 groups) $[\{\sigma_8 \ \sigma_{16}\}_{G7}, \{\sigma_9 \ \sigma_{24}\}_{G8}, \{\sigma_{13} \ \sigma_{10}\}_{G9}, \{\sigma_{15} \ \sigma_{23}\}_{G10}, \{\sigma_{19} \ \sigma_{12}\}_{G11}, \{\sigma_{20} \ \sigma_{17}\}_{G12}]$

Torsion modes (3 groups): $[\{\sigma_{11}\}_{G13}, \{\sigma_{18}\}_{G14}, \{\sigma_{22}\}_{G15}]$

Saddle modes (3 groups): $[\{\sigma_{29} \ \sigma_{30} \ \sigma_{38}\}_{G16}, \{\sigma_{28} \ \sigma_{33} \ \sigma_{36}\}_{G17}, \{\sigma_{27} \ \sigma_{34} \ \sigma_{39}\}_{G18}]$

Zero-energy stress modes (1 group): $[\{\sigma_7, \sigma_{14}, \sigma_{21}, \sigma_{25}, \sigma_{26}, \sigma_{31}, \sigma_{32}, \sigma_{35}, \sigma_{37}\}_{G19}]$

The first 18 stress mode groups correspond to (m=n-r=) 18 natural deformation modes and the last group corresponds to rigid body modes. Similar to the 2-D case, there are many other ways to derive initial stress modes. For example, in the assumed stress matrix presented in ref. [3.5], there are 12 stress modes that are different from stress modes $\{\sigma_1\}$-$\{\sigma_{39}\}$. These stress modes can be expressed as follows,

Tension and compression modes,

$$\{\sigma_{40}\}=\begin{Bmatrix}1\\1\\1\\0\\0\\0\end{Bmatrix} \quad \{\sigma_{41}\}=\begin{Bmatrix}1\\-1\\0\\0\\0\\0\end{Bmatrix} \quad \{\sigma_{42}\}=\begin{Bmatrix}-1\\-1\\2\\0\\0\\0\end{Bmatrix} \tag{3-144}$$

Symmetric bending modes,

$$\{\sigma_{43}\}=\begin{Bmatrix}z\\z\\0\\0\\0\\0\end{Bmatrix} \quad \{\sigma_{44}\}=\begin{Bmatrix}0\\x\\x\\0\\0\\0\end{Bmatrix} \quad \{\sigma_{45}\}=\begin{Bmatrix}y\\0\\y\\0\\0\\0\end{Bmatrix} \tag{3-145}$$

Anti-symmetric bending modes,

$$\{\sigma_{46}\}=\begin{Bmatrix}z\\-z\\0\\0\\0\\0\end{Bmatrix} \quad \{\sigma_{47}\}=\begin{Bmatrix}0\\x\\-x\\0\\0\\0\end{Bmatrix} \quad \{\sigma_{48}\}=\begin{Bmatrix}y\\0\\-y\\0\\0\\0\end{Bmatrix} \tag{3-146}$$

Torsion modes,

$$\{\sigma_{49}\}=\begin{Bmatrix}0\\0\\0\\z\\x\\y\end{Bmatrix} \quad \{\sigma_{50}\}=\begin{Bmatrix}0\\0\\0\\z\\-x\\0\end{Bmatrix} \quad \{\sigma_{51}\}=\begin{Bmatrix}0\\0\\0\\z\\x\\-2y\end{Bmatrix} \tag{3-147}$$

In the stress matrices derived by means of the symmetric group theory[3.34,3.35], there are eleven stress modes that are different from stress modes $\{\sigma_1\}$-$\{\sigma_{51}\}$. They can be expressed in the form,

Tension and compressive mode,

$$\{\sigma_{52}\}=\begin{Bmatrix} 0 \\ 1 \\ -1 \\ 0 \\ 0 \\ 0 \end{Bmatrix} \tag{3-148}$$

Torsion modes,

$$\{\sigma_{53}\}=\begin{Bmatrix} 0 \\ 0 \\ 0 \\ 0 \\ -x \\ y \end{Bmatrix} \tag{3-149}$$

Bending modes,

$$\{\sigma_{54}\ \sigma_{55}\ \sigma_{56}\}=\begin{Bmatrix} 2x & 0 & 0 \\ 0 & 2y & 0 \\ 0 & 0 & 2z \\ -y & -x & 0 \\ 0 & -z & -y \\ -z & 0 & -x \end{Bmatrix} \tag{3-150}$$

and

$$\{\sigma_{57}\ \sigma_{58}\ \sigma_{59}\}=\begin{Bmatrix} 0 & 0 & 0 \\ 0 & 0 & 0 \\ 0 & 0 & 0 \\ y & x & 0 \\ 0 & -z & -y \\ -z & 0 & x \end{Bmatrix} \tag{3-151}$$

Saddle modes,

$$\{\sigma_{60}\ \sigma_{61}\ \sigma_{62}\}=\begin{Bmatrix} 0 & 0 & 0 \\ 0 & 0 & 0 \\ 0 & 0 & 0 \\ -2xz & -2yz & x^2+y^2 \\ y^2+z^2 & -2xy & -2xz \\ -2xy & x^2+z^2 & -2yz \end{Bmatrix} \tag{3-152}$$

Other stress modes may be also needed in an assumed stress matrix in order to describe special stress distribution in a local region of a structure to be analyzed. For instance,

Bending modes,

$$\{\sigma_{63}\}=\begin{Bmatrix} z \\ z \\ z \\ 0 \\ 0 \\ 0 \end{Bmatrix} \quad \{\sigma_{64}\}=\begin{Bmatrix} x \\ x \\ x \\ 0 \\ 0 \\ 0 \end{Bmatrix} \quad \{\sigma_{65}\}=\begin{Bmatrix} y \\ y \\ y \\ 0 \\ 0 \\ 0 \end{Bmatrix} \tag{3-153}$$

Saddle modes,

$$\{\sigma_{66}\}=\begin{Bmatrix} yz \\ yz \\ yz \\ 0 \\ 0 \\ 0 \end{Bmatrix} \quad \{\sigma_{67}\}=\begin{Bmatrix} zx \\ zx \\ zx \\ 0 \\ 0 \\ 0 \end{Bmatrix} \quad \{\sigma_{68}\}=\begin{Bmatrix} xy \\ xy \\ xy \\ 0 \\ 0 \\ 0 \end{Bmatrix} \tag{3-154}$$

Tension and compression mode,

$$\{\sigma_{69}\}=\begin{Bmatrix} z^2 \\ z^2 \\ 0 \\ 0 \\ 0 \\ 0 \end{Bmatrix} \tag{3-155}$$

According to the steps 13 -14 in the proposed procedure of classification, the 30 new stress modes $\{\sigma_{40}\}$-$\{\sigma_{69}\}$ can be classified into different stress mode groups above as follows,

Tension and compression modes (3 groups): $[\{\sigma_1, \sigma_{40}, \sigma_{69}\}_{G1}, \{\sigma_2, \sigma_{41}\}_{G2}, \sigma_3, \sigma_{42}, \sigma_{52}\}_{G3}]$

Pure shear modes (3 groups): $[\{\sigma_4\}_{G4}, \{\sigma_5\}_{G5}, \{\sigma_6\}_{G6}]$

Bending modes (6 groups): $[\{\sigma_8, \sigma_{16}, \sigma_{44}, \sigma_{54}, \sigma_{64}\}_{G7}, \{\sigma_9, \sigma_{24}, \sigma_{47}, \sigma_{57}\}_{G8}, \{\sigma_{13}, \sigma_{10}, \sigma_{45}, \sigma_{55}, \sigma_{65}\}_{G9}, \{\sigma_{15}, \sigma_{23}, \sigma_{48}, \sigma_{58}\}_{G10}, \{\sigma_{19}, \sigma_{12}, \sigma_{43}, \sigma_{56}, \sigma_{63}\}_{G11}, \{\sigma_{20}, \sigma_{17}, \sigma_{46}, \sigma_{59}\}_{G12}]$

Torsion modes (3 groups): $[\{\sigma_{11}, \sigma_{49}\}_{G13}, \{\sigma_{18}, \sigma_{50}\}_{G14}, \{\sigma_{22}, \sigma_{51}, \sigma_{53}\}_{G15}]$

Saddle modes (3 groups): $[\{\sigma_{29}, \sigma_{30}, \sigma_{38}, \sigma_{66}, \sigma_{60}\}_{G16}, \{\sigma_{28}, \sigma_{33}, \sigma_{36}, \sigma_{67}, \sigma_{61}\}_{G17}, \{\sigma_{27}, \sigma_{34}, \sigma_{39}, \sigma_{68}, \sigma_{62}\}_{G18}]$

Zero-energy stress modes (1 group): $[\{\sigma_7, \sigma_{14}, \sigma_{21}, \sigma_{25}, \sigma_{26}, \sigma_{31}, \sigma_{32}, \sigma_{35}, \sigma_{37}\}_{G19}]$

More stress modes can be classified into the stress mode groups above. If the flexibility matrix [H] is a diagonal matrix, the stress modes are uncoupled and the classification of the stress modes is unique (see theorem 3). Otherwise, some stress modes may appear in more than one group.

3.4.4 Construction of Assumed Stress Matrix

As shown above, by means of the proposed procedure for the classification of stress mode, stress modes can be classified into m (=n-r) stress mode groups corresponding to m natural deformation modes and a zero energy mode group corresponding to rigid body modes. Each natural deformation mode is related to a stress mode group except zero energy mode group, and each stress mode group may contain many different stress modes that are interchangeable in the stress matrix [P]. Thus, based on the iso-function stress matrix and classified stress mode groups, the method can be established for determining the assumed stress matrix of a hybrid element.

Assumed Stress Matrix of Hybrid Element

The classification of stress modes reveals the relationship among the different stress modes that are used in the different stress matrices for any type of hybrid element proposed by different researchers. In order to avoid kinematic deformation mode, the stress matrix [P] must contain m stress modes at least. No matter how many stress modes there are in the stress matrix [P], the order of the stiffness matrix is equal to or less than m. Therefore, m stress modes is necessary and sufficient to form a stress matrix for avoiding kinematic deformation modes in the hybrid element. Moreover, in view of the classification of stress modes, the m stress modes in the stress matrix [P] must come from m different stress mode groups except zero energy mode group. Thus, for a hybrid element to be free from kinematic deformation mode,

one has the necessary and sufficient condition:

The number of stress modes in an assumed stress matrix must be equal to or more than m (= n-r) and at least m stress modes in the stress matrix [P] must be chosen from m different stress mode groups corresponding to m natural deformation modes of an element which has n degrees of freedom and r rigid body modes.

In this statement, the necessary condition is that the number of stress modes for a hybrid element must be equal to or more than m (=n-r). It was presented by F. Veubeke [3.36] and Pian [3.37]. The sufficient condition is that the stress matrix [P] must contain m stress modes chosen from m different stress mode groups corresponding to m natural deformation modes. This condition explains why in some examples there exist kinematic deformation modes even when the necessary condition (m' > n-r) is satisfied. In these examples, the stress modes in the stress matrix [P] do not come from m different stress mode groups except the zero energy mode group.

For a hybrid element, an assumed stress field, its stress matrix contains m (=n-r) least-order stress modes and its resulting finite element is free from kinematic deformation modes, is considered to be best and is optimal with respect to computer resources [3.38,3.34] because overuse of stress modes will result in over-rigid element [3.38], and the calculation of element stiffness matrix requires an inversion of the flexibility matrix [H]. By means of the m classified stress mode groups and the necessary and sufficient condition, this kind of stress matrices can be constructed. Furthermore, it is convenient to construct an assumed stress matrix according to the problem to be solved because there are many stress modes in every stress mode group for choice. The procedure of constructing stress matrix is presented as follows,

Step 1

Using the iso-function method, one can derive a number of initial stress modes to be classified.

Step 2

One may put the initial stress modes one by one into stress matrix [P] in the order from low order term to high order term. By means of the classification method, one can obtain m representative stress modes corresponding to m natural deformation modes. These representative stress modes form a optimal stress matrix $[P]_{opt}$ from the existing stress matrix $[P]_{iso}$.

Step 3

One may obtain other initial stress modes derived by different methods. Following the steps 13 - 14 in the procedure of the classification, one can classify all initial stress modes into m+1 different stress mode groups.

Step 4

By means of the m+1 classified stress mode groups and the necessary and sufficient condition above, many stress matrices [P] can be constructed according to the problem to be solved. It is necessary to choose one stress mode at least from each group except the zero energy mode group in order to avoid kinematic deformation modes.

The necessary steps have been illustrated in the section above. The following gives some examples to illustrate the procedure for constructing a stress matrix [P] which has minimum number of stress modes.

2-D, 4-node plane hybrid element

By means of the m+1 stress mode groups classified above and the necessary and sufficient condition for avoiding kinematic deformation modes, one can choose one stress mode from each stress mode group except zero energy mode group to form a stress matrix. For example,

$$[P_{III}] = [\boldsymbol{\sigma}_{10}\ \boldsymbol{\sigma}_{11}\ \boldsymbol{\sigma}_3\ \boldsymbol{\sigma}_{12}\ \boldsymbol{\sigma}_{13}] = \begin{bmatrix} 1 & 1 & 0 & 0 & -x \\ 1 & -1 & 0 & -y & 0 \\ 0 & 0 & 1 & x & y \end{bmatrix} \tag{3-156}$$

and

$$[P_{IV}] = [\boldsymbol{\sigma}_{10}\ \boldsymbol{\sigma}_{11}\ \boldsymbol{\sigma}_3\ \boldsymbol{\sigma}_7\ \boldsymbol{\sigma}_5] = \begin{bmatrix} 1 & 1 & 0 & y & 0 \\ 1 & -1 & 0 & 0 & x \\ 0 & 0 & 1 & 0 & 0 \end{bmatrix} \tag{3-157}$$

Five stress modes in each stress matrix come from five different stress mode groups corresponding to five natural deformation modes. The two stress matrices are the same as that proposed by Atluri[3.34,3.35]. More stress matrices can be constructed on purpose. For example,

$$[P_V] = [\sigma_{10}\ \sigma_{11}\ \sigma_3\ \sigma_7\ \sigma_{13}] = \begin{bmatrix} 1 & 1 & 0 & y & -x \\ 1 & -1 & 0 & 0 & 0 \\ 0 & 0 & 1 & 0 & y \end{bmatrix} \tag{3-158}$$

and

$$[P_{VI}] = [\sigma_{14}\ \sigma_{11}\ \sigma_3\ \sigma_{12}\ \sigma_{13}] = \begin{bmatrix} x^2 & 1 & 0 & 0 & -x \\ 0 & -1 & 0 & -y & 0 \\ 0 & 0 & 1 & x & y \end{bmatrix} \tag{3-159}$$

The eigenvalue examination shows that the hybrid element constructed by $[P_I]$ - $[P_{VI}]$ are free from kinematic deformation modes as shown in Table 2. In the last column of the table, the eigenvalues of displacement element stiffness matrix are given. If stress modes in a stress matrix [P] come from m_1 (<m) stress mode groups, the hybrid element will have kinematic deformation modes even if the number of stress modes is larger than m. This is why a hybrid element contains kinematic deformation modes when the necessary condition (m' > n-r) is satisfied. A stress matrix [P] must have m stress modes corresponding to m natural deformation modes of an element.

Table 2 Eigenvalues of stiffness matrices (2-D, 4-node plane element, ν=0.3)

$[P_I]$	$[P_{II}]$	$[P_{III}]$	$[P_{IV}]$	$[P_V]$	$[P_{VI}]$	Disp.
0.4945	0.3333	0.09259	0.3333	0.09259	0.09259	0.4945
0.4945	0.3333	0.09259	0.3333	0.3333	0.09259	0.4945
0.7692	0.7692	0.7692	0.7692	0.7692	0.7692	0.7692
0.7692	0.7692	0.7692	0.7692	0.7692	0.7692	0.7692
1.4290	1.4290	1.4290	1.4290	1.4290	1.4290	1.4290

3-D, 8-node solid hybrid element

Using the same way as the 2-D case, one can choose m stress modes from m classified stress mode groups except zero energy mode group above to form the eight stress matrices $[P_2]$ - $[P_9]$ proposed by Atluri et al [3.34-3.35] as follows,

$$[P_2] = [\sigma_{40}\ \sigma_{41}\ \sigma_{42}\ \sigma_4\ \sigma_5\ \sigma_6\ \sigma_{49}\ \sigma_{50}\ \sigma_{53}\ \sigma_{54}\ \sigma_{55}\ \sigma_{56}\ \sigma_{57}\ \sigma_{58}\ \sigma_{59}\ \sigma_{60}\ \sigma_{61}\ \sigma_{62}]$$

$$= \begin{bmatrix} 1 & 1 & 0 & 0 & 0 & 0 & 0 & 0 & 0 & 2x & 0 & 0 & 0 & 0 & 0 & 0 & 0 & 0 \\ 1 & -1 & 1 & 0 & 0 & 0 & 0 & 0 & 0 & 0 & 2y & 0 & 0 & 0 & 0 & 0 & 0 & 0 \\ 1 & 0 & -1 & 0 & 0 & 0 & 0 & 0 & 0 & 0 & 0 & 2z & 0 & 0 & 0 & 0 & 0 & 0 \\ 0 & 0 & 0 & 1 & 0 & 0 & z & z & 0 & -y & -x & 0 & y & x & 0 & -2xz & -2yz & x^2+y^2 \\ 0 & 0 & 0 & 0 & 1 & 0 & x & -x & -x & 0 & -z & -y & 0 & -z & -y & y^2+z^2 & -2xy & -2xz \\ 0 & 0 & 0 & 0 & 0 & 1 & y & 0 & y & -z & 0 & -x & -z & 0 & x & -2xy & x^2+z^2 & -2yz \end{bmatrix}$$

(3-160)

$$[P_3] = [\sigma_{40}\ \sigma_{41}\ \sigma_{42}\ \sigma_4\ \sigma_5\ \sigma_6\ \sigma_{49}\ \sigma_{50}\ \sigma_{53}\ \sigma_{54}\ \sigma_{55}\ \sigma_{56}\ \sigma_{57}\ \sigma_{58}\ \sigma_{59}\ \sigma_{27}\ \sigma_{36}\ \sigma_{30}]$$

$$= \begin{bmatrix} 1 & 1 & 0 & 0 & 0 & 0 & 0 & 0 & 0 & 2x & 0 & 0 & 0 & 0 & 0 & 0 & 0 & yz \\ 1 & -1 & 1 & 0 & 0 & 0 & 0 & 0 & 0 & 0 & 2y & 0 & 0 & 0 & 0 & 0 & xz & 0 \\ 1 & 0 & -1 & 0 & 0 & 0 & 0 & 0 & 0 & 0 & 0 & 2z & 0 & 0 & 0 & xy & 0 & 0 \\ 0 & 0 & 0 & 1 & 0 & 0 & z & z & 0 & -y & -x & 0 & y & x & 0 & 0 & 0 & 0 \\ 0 & 0 & 0 & 0 & 1 & 0 & x & -x & -x & 0 & -z & -y & 0 & -z & -y & 0 & 0 & 0 \\ 0 & 0 & 0 & 0 & 0 & 1 & y & 0 & y & -z & 0 & -x & -z & 0 & x & 0 & 0 & 0 \end{bmatrix}$$

(3-161)

and

$$[P_4] = [\sigma_{40}\ \sigma_{41}\ \sigma_{42}\ \sigma_4\ \sigma_5\ \sigma_6\ \sigma_{49}\ \sigma_{50}\ \sigma_{53}\ \sigma_{54}\ \sigma_{55}\ \sigma_{56}\ \sigma_{48}\ \sigma_{47}\ \sigma_{46}\ \sigma_{60}\ \sigma_{61}\ \sigma_{62}]$$

$$[P_5] = [\sigma_{40}\ \sigma_{41}\ \sigma_{42}\ \sigma_4\ \sigma_5\ \sigma_6\ \sigma_{49}\ \sigma_{50}\ \sigma_{53}\ \sigma_{54}\ \sigma_{55}\ \sigma_{56}\ \sigma_{48}\ \sigma_{47}\ \sigma_{46}\ \sigma_{27}\ \sigma_{36}\ \sigma_{30}]$$

$$[P_6] = [\sigma_{40}\ \sigma_{41}\ \sigma_{42}\ \sigma_4\ \sigma_5\ \sigma_6\ \sigma_{49}\ \sigma_{50}\ \sigma_{53}\ \sigma_{45}\ \sigma_{44}\ \sigma_{43}\ \sigma_{57}\ \sigma_{58}\ \sigma_{59}\ \sigma_{60}\ \sigma_{61}\ \sigma_{62}]$$

$$[P_7] = [\sigma_{40}\ \sigma_{41}\ \sigma_{42}\ \sigma_4\ \sigma_5\ \sigma_6\ \sigma_{49}\ \sigma_{50}\ \sigma_{53}\ \sigma_{45}\ \sigma_{44}\ \sigma_{43}\ \sigma_{57}\ \sigma_{58}\ \sigma_{59}\ \sigma_{27}\ \sigma_{36}\ \sigma_{30}]$$

$$[P_8] = [\sigma_{40}\ \sigma_{41}\ \sigma_{42}\ \sigma_4\ \sigma_5\ \sigma_6\ \sigma_{49}\ \sigma_{50}\ \sigma_{53}\ \sigma_{45}\ \sigma_{44}\ \sigma_{43}\ \sigma_{48}\ \sigma_{47}\ \sigma_{46}\ \sigma_{60}\ \sigma_{61}\ \sigma_{62}]$$

$$[P_9] = [\sigma_{40}\ \sigma_{41}\ \sigma_{42}\ \sigma_4\ \sigma_5\ \sigma_6\ \sigma_{49}\ \sigma_{50}\ \sigma_{53}\ \sigma_{45}\ \sigma_{44}\ \sigma_{43}\ \sigma_{48}\ \sigma_{47}\ \sigma_{46}\ \sigma_{27}\ \sigma_{36}\ \sigma_{30}]$$

(3-162)

The assumed stress matrix given by Huang [3.5] also can be formed the same way,

$$[P_{10}] = [\sigma_{40}\ \sigma_{41}\ \sigma_{42}\ \sigma_{4}\ \sigma_{5}\ \sigma_{6}\ \sigma_{43}\ \sigma_{44}\ \sigma_{45}\ \sigma_{46}\ \sigma_{47}\ \sigma_{48}\ \sigma_{49}\ \sigma_{50}\ \sigma_{51}\ \sigma_{27}\ \sigma_{36}\ \sigma_{30}]$$

$$= \begin{bmatrix} 1 & 1 & -1 & 0 & 0 & 0 & z & 0 & y & z & 0 & y & 0 & 0 & 0 & yz & 0 & 0 \\ 1 & -1 & -1 & 0 & 0 & 0 & z & x & 0 & -z & x & 0 & 0 & 0 & 0 & 0 & zx & 0 \\ 1 & 0 & 2 & 0 & 0 & 0 & 0 & x & y & 0 & -x & -y & 0 & 0 & 0 & 0 & 0 & xy \\ 0 & 0 & 0 & 1 & 0 & 0 & 0 & 0 & 0 & 0 & 0 & 0 & z & z & z & 0 & 0 & 0 \\ 0 & 0 & 0 & 0 & 1 & 0 & 0 & 0 & 0 & 0 & 0 & 0 & x & -x & -x & 0 & 0 & 0 \\ 0 & 0 & 0 & 0 & 0 & 1 & 0 & 0 & 0 & 0 & 0 & 0 & y & 0 & -2y & 0 & 0 & 0 \end{bmatrix} \tag{3-163}$$

Moreover, many stress matrices [P] can be constructed on purpose. Three new stress matrices are given as follows,

$$[P_1^*] = [\sigma_{40}\ \sigma_{41}\ \sigma_{42}\ \sigma_{4}\ \sigma_{5}\ \sigma_{6}\ \sigma_{63}\ \sigma_{64}\ \sigma_{65}\ \sigma_{46}\ \sigma_{47}\ \sigma_{48}\ \sigma_{49}\ \sigma_{50}\ \sigma_{51}\ \sigma_{66}\ \sigma_{67}\ \sigma_{68}]$$

$$= \begin{bmatrix} 1 & 1 & -1 & 0 & 0 & 0 & z & x & y & z & 0 & y & 0 & 0 & 0 & yz & xz & xy \\ 1 & -1 & -1 & 0 & 0 & 0 & z & x & y & -z & x & 0 & 0 & 0 & 0 & yz & xz & xy \\ 1 & 0 & 2 & 0 & 0 & 0 & z & x & y & 0 & -x & -y & 0 & 0 & 0 & yz & xz & xy \\ 0 & 0 & 0 & 1 & 0 & 0 & 0 & 0 & 0 & 0 & 0 & 0 & z & z & z & 0 & 0 & 0 \\ 0 & 0 & 0 & 0 & 1 & 0 & 0 & 0 & 0 & 0 & 0 & 0 & x & -x & -x & 0 & 0 & 0 \\ 0 & 0 & 0 & 0 & 0 & 1 & 0 & 0 & 0 & 0 & 0 & 0 & y & 0 & -2y & 0 & 0 & 0 \end{bmatrix} \tag{3-164}$$

$$[P_2^*] = [\sigma_{40}\ \sigma_{41}\ \sigma_{42}\ \sigma_{4}\ \sigma_{5}\ \sigma_{6}\ \sigma_{43}\ \sigma_{44}\ \sigma_{45}\ \sigma_{57}\ \sigma_{58}\ \sigma_{59}\ \sigma_{22}\ \sigma_{11}\ \sigma_{18}\ \sigma_{30}\ \sigma_{36}\ \sigma_{27}]$$

$$= \begin{bmatrix} 1 & 0 & 0 & 0 & 0 & 0 & z & 0 & y & 0 & 0 & 0 & 0 & 0 & 0 & yz & 0 & 0 \\ 0 & 1 & 0 & 0 & 0 & 0 & z & x & 0 & 0 & 0 & 0 & 0 & 0 & 0 & 0 & xz & 0 \\ 0 & 0 & 1 & 0 & 0 & 0 & 0 & x & y & 0 & 0 & 0 & 0 & 0 & 0 & 0 & 0 & xy \\ 0 & 0 & 0 & 1 & 0 & 0 & 0 & 0 & 0 & y & x & 0 & z & 0 & 0 & 0 & 0 & 0 \\ 0 & 0 & 0 & 0 & 1 & 0 & 0 & 0 & 0 & 0 & -z & -y & 0 & x & 0 & 0 & 0 & 0 \\ 0 & 0 & 0 & 0 & 0 & 1 & 0 & 0 & 0 & -z & 0 & x & 0 & 0 & y & 0 & 0 & 0 \end{bmatrix} \tag{3-165}$$

$$
[P_3^*] = [\sigma_{69}\ \sigma_{41}\ \sigma_{42}\ \sigma_4\ \sigma_5\ \sigma_6\ \sigma_{63}\ \sigma_{64}\ \sigma_{65}\ \sigma_{46}\ \sigma_{47}\ \sigma_{48}\ \sigma_{49}\ \sigma_{50}\ \sigma_{51}\ \sigma_{66}\ \sigma_{67}\ \sigma_{68}]
$$

$$
= \begin{bmatrix}
z^2 & 1 & -1 & 0 & 0 & 0 & z & x & y & z & 0 & y & 0 & 0 & 0 & yz & xz & xy \\
z^2 & -1 & -1 & 0 & 0 & 0 & z & x & y & -z & x & 0 & 0 & 0 & 0 & yz & xz & xy \\
0 & 0 & 2 & 0 & 0 & 0 & z & x & y & 0 & -x & -y & 0 & 0 & 0 & yz & xz & xy \\
0 & 0 & 0 & 1 & 0 & 0 & 0 & 0 & 0 & 0 & 0 & 0 & z & z & z & 0 & 0 & 0 \\
0 & 0 & 0 & 0 & 1 & 0 & 0 & 0 & 0 & 0 & 0 & 0 & x & -x & -x & 0 & 0 & 0 \\
0 & 0 & 0 & 0 & 0 & 1 & 0 & 0 & 0 & 0 & 0 & 0 & y & 0 & -2y & 0 & 0 & 0
\end{bmatrix}
$$

(1-166)

The results of eigenvalue examination are given in Table 3. It shows that each of the stiffness matrices constructed by the assumed stress matrices $[P_1]$ - $[P_{10}]$, $[P_1*]$, $[P_2*]$, and $[P_3*]$ has m non-zero eigenvalues. The resulting hybrid elements do not have any kinematic deformation modes.

More assumed stress matrices can also be constructed by means of this method. If one stress mode group is missed except the zero energy mode group in the process of choosing stress modes, the hybrid element will contain kinematic deformation modes. In the previous work, it is proposed to suppress kinematic deformation modes by adding stress modes of high order term. Actually, it can not guarantee that all kinematic deformation modes are suppressed. If the high order stress modes do not belong to the stress mode groups which are missed in the construction of the assumed stress matrix except the zero energy mode group, adding stress modes of high order term can not improve the hybrid element any more. Moreover, overuse of stress modes will result in over-rigid elements[3.38].

Table 3 Eigenvalues of stiffness matrices (3-D, 8-node solid element, ν=0.3)

$[P_2]$	$[P_4]$	$[P_1]$, $[P_9]$, $[P_{10}]$	$[P_7]$, $[P_2^*]$	$[P_1^*]$	$[P_3^*]$
0.07123	0.07123	0.1111	0.1111	0.09259	0.09259
0.07123	0.07123	0.1111	0.1111	0.09259	0.09259
0.07123	0.07123	0.1111	0.1111	0.09259	0.09259
0.1282	0.2564	0.2564	0.1282	0.2564	0.2564
0.1282	0.2564	0.2564	0.1282	0.2564	0.2564
0.1282	0.2564	0.2564	0.1282	0.2564	0.2564
0.1282	0.1282	0.1282	0.1282	0.1282	0.1282
0.1282	0.1282	0.1282	0.1282	0.1282	0.1282
0.07246	0.07264	0.4762	0.4762	0.5556	0.5556
0.07246	0.07264	0.4762	0.4762	0.5556	0.5556
0.07246	0.07264	0.4762	0.4762	0.5556	0.5556
0.5128	0.5128	0.5128	0.5128	0.5128	0.5128
0.7692	0.7692	0.7692	0.7692	0.7692	0.7692
0.7692	0.7692	0.7692	0.7692	0.7692	0.7692
0.7692	0.7692	0.7692	0.7692	0.7692	0.7692
0.7692	0.7692	0.7692	0.7692	0.7692	0.7692
0.7692	0.7692	0.7692	0.7692	0.7692	0.7692
2.5000	2.5000	2.5000	2.5000	2.5000	0.8065

Therefore, an ideal situation is to choose m (=n-r) least-order stress modes, but with the suppression of all kinematic deformation modes. Thus, an assumed stress matrix [P] can be constructed by choosing m stress modes from m stress mode groups that correspond to m natural deformation modes.

Optimal Stress Matrix for Partial Hybrid Element

The classification method can be used to determine an optimal stress matrix for a hybrid element, and it is also available to determine an optimal partial stress matrix for a partial hybrid element. The difference is the number of stress mode groups. For partial hybrid element, the number of stress modes groups is equal to

$$m=n-r-n_d \tag{3-167}$$

Where n and r is the same as that in conventional hybrid elements. In section 3.3, it has been shown that the stiffness matrix of a partial hybrid element (3-35) consists of two parts: the displacement-formulated stiffness matrix and the hybrid-formulated stiffness matrix. In equation (3-167), n_d is the rank of the displacement-formulated stiffness matrix. Therefore, it is necessary to calculate the rank of the displacement-formulated stiffness matrix to determine an optimal stress matrix for a partial hybrid element. The necessary and sufficient condition for partial hybrid elements becomes

The necessary and sufficient condition.

The number of stress modes in an assumed partial stress matrix must be equal to or more than m (= n-r-n_d) and at least m stress modes in the partial stress matrix [P] must be chosen from m different stress mode groups corresponding to m natural deformation modes of an element which has n degrees of freedom, r rigid body modes, and n_d order displacement-formulated stiffness matrix .

The procedure to construct an optimal partial stress matrix for a partial hybrid element becomes

Step 1

Examine the rank of the displacement-formulated stiffness matrix of a partial hybrid element.

Step 2

Using the iso-function method, one can derive a number of initial stress modes to be classified.

Step 3

One may put the initial stress modes one by one into partial stress matrix [P] in the order from low order term to high order term. By means of the classification method, one can obtain m representative stress modes corresponding to m natural deformation modes. These representative stress modes form an optimal partial stress matrix $[P_g]_{opt}$ from the existing partial stress matrix $[P_g]_{iso}$.

Following this procedure, an optimal assumed partial stress field can be determined. Some examples will be given in chapter 4.

The classification method presented in this chapter can be applied to any type of hybrid elements. Usually, it is used for two purposes:

1. Determine the optimal stress matrix from the existing stress matrix $[P]_{iso}$ or any other stress matrix [P] derived using other methods, and classify stress modes into m different stress mode groups;

2. Construct many new assumed stress matrices by using minimum number of stress modes according to the problems to be analyzed. These stress matrices are without zero-energy stress modes, and the resulting element stiffness matrices are free from kinematic deformation modes.

The classification of stress modes reveals the relationship among the different assumed stress fields for any type of hybrid element proposed by different researchers. An assumed stress matrix [P], which consists of m (=n-r) least-order stress modes and results in the element stiffness matrix without kinematic deformation modes, is considered to be best and is optimal with respect to computer resources.

REFERENCES

3.1 S. V. Hoa and W. Feng, 'Finite Elements for Analysis of Composite Structures', in Ed., S.V. Hoa, Computer-Aided Design of Polymer-Matrix Composite Structures, Marcel Dekker, Inc., 1995.

3.2 R. L. Spilker, 'Hybrid-stress eight-node elements for thin and thick multilayer laminated plates', Int. J. Numer. Methods Engrg, vol.18, 801-828(1982).

3.3 W. J. Liou & C.T. Sun, 'A three dimensional hybrid stress isoparametric element for the analysis of laminated composite plates', Computers & Structures, vol. 25, no. 2, 241-249 (1987).

3.4 Q. Huang, S.V. Hoa and T.S. Sankar, 'Three Dimensional Finite Element Formulation for Stress Analysis of Anisotropic Laminated Structures', in Eds., W.P.de Wilde and W.R. Blain, Composite Material Design and Analysis, Computational Mechanics Publisher, 1990.

3.5 Q. Huang, 'Three Dimensional Composite Finite Element for Stress Analysis of Anisotropic Laminate Structures', Ph. D. Dissertation, Concordia University, Montreal, Canada (1989).

3.6 J. Han, 'Three dimensional multilayer composite finite element for stress analysis of composite laminates', Ph.D. Dissertation, Concordia University, Montreal, Canada, (1994).

3.7 J. Han. and S.V. Hoa, 'A three-dimensional multilayer composite finite element for stress analysis of composite laminates', Int. J. Numer. Methods Engrg., vol.36, 3903- 3914(1993).

3.8 W. Feng and S.V. Hoa, 'A partial hybrid degenerated plate/shell element for the analysis of laminated composites', Int. J. Numer. Methods Eng., vol. 39, 3625-3639(1996).

3.9 W. Feng and S.V. Hoa, '3-D transition element formulation for the global-local analysis of laminated structures', Int. Conf. on Design and Manufacturing Using Composites, Montreal, Canada (1994).

3.10 W. Feng and S.V. Hoa, 'A 3-D partial hybrid laminated element for analysis of thick laminates', Third Int. Conf. on Composites Engineering, New Orleans, USA (1996).

3.11 W. Feng and S.V. Hoa, 'A multilayer element with partial assumed stress field for analysis of

laminated structures', The 16th Canadian Congress of Applied Mechanics CANCAM 97, Quebec, Canada (1997).
3.12 V. Hoa and W. Feng, 'Application of a global/local finite element model to composite laminates', Science and Engineering of Composite Materials, vol. 5, 151-168(1996).
3.13 T. H. H. Pian and M.-S. Li, 'Stress analysis of laminated composites by hybrid finite elements', in Discretization Methods in Structural Mechanics (Ed. Kuhn, G. and Mang, H.), 1989.
3.14 H. S. Jing & M.-L. Liao, 'Partial hybrid stress element for the analysis of thick laminated composite plates', Int. J. Numer. Methods Eng., vol. 28, 2813-2827(1989).
3.15 C. Y. Wang and H.-K. Ching, 'A modified partial hybrid stress finite element method for the aminated composite plate analysis', Composites Modelling and Processing Science, ICCM/9, vol. III, ed., Antonio Miravete, Woodhead Publishing Limited(1993).
3.16 W. Feng, S.V. Hoa, and Q. Huang, 'Classification of stress modes in assumed stress fields of hybrid finite elements', Int. J. for Numer. Methods in Eng., (Accepted).
3.17 E. Reissner, 'On a certain mixed variational theorem and a proposed application', Int. J. Numer. Methods Eng., vol. 20, 1366-1368(1984).
3.18 E. Reissner, 'On a mixed variational theorem and on shear deformable plate theory', Int. J. Numer. Methods Eng., vol. 23, 193-198(1986).
3.19 K. Moriya, 'Laminated plate and shell elements for finite element analysis of advanced fibre reinforced composite structures (Japanese), Trans. Japan Soc. Mech. Eng. (series A), 52, no. 478, 1600-1607(1986).
3.20 S. Ahmad and B.M. Irons, 'An assumed stress approach to refined isoparametric finite elements in three dimensions', Finite Element Method in Engineering, University of New South Wales, 85-100(1974).
3.21 R. Rubinstein, E.F. Punch and S.N. Atluri, Computer Methods in Applied Mechanics and Engineering, vol. 38, 63-92(1983).
3.22 T.H.H. Pian and D.P. Chen, ' On the suppression of zero energy deformation modes', International Journal for Numerical Methods in Engineering, vol. 19, 1741-1752(1983).
3.23 T.H.H. Pian and P. Tong, ' Relations between incompatible displacement model and hybrid stress model', Int. J. Numer. Methods Eng., vol.22, 173-181(1986).
3.24 T.H.H. Pian and C.C. Wu, ' A rational approach for choosing stress terms for hybrid finite element formulations', Int. J. Numer. Methods Eng., vol.26, 2331-2343(1988).
3.25 K.Y. Sze, C.L. Chow and W.J. Chen, ' A rational formulation of iso-parametric hybrid stress element for three dimensional stress analysis', Finite Elements Analysis and Design, vol. 7, 61-72(1990).
3.26 W.J. Chen and Y.K. Cheung, ' Three-dimensional 8-node and 20-node refined hybrid isoparametric elements', International Journal for Numerical Methods in Engineering, vol.35, 1871-1889(1992).
3.27 K.Y. Sze, ' Control of spurious mechanisms for 20-node and transition sub-integrated hexahedral elements', International Journal for Numerical Methods in Engineering, vol.37, 2235-2250(1994).
3.28 C.C. Wu and Y.K. Cheung, 'On optimization approaches of hybrid stress elements', Finite Elements in Analysis and Design, vol. 21, 111-128(1995).
3.29 Q. Huang, Modal Analysis of Deformable Bodies with Finite degrees of Deformation Freedom, Shanghai Institute of Applied Mathematics and Mechanics, Shanghai University, Technical Report, 1990.
3.30 B. Fraeijs de Veubeke, 'Displacement and equilibrium models in the finite element methods', in: O. C. Zienkiewicz and G. S. Holister, (eds.), Stress Analysis, John Wiley, London, 1965.
3.31 H. Stolarski and T. Belytschko, 'Limitation principles for mixed finite elements based on the Hu-Washizu variational formulation', Comput. Meth. Appl. Mech. Engng. vol. 60, 195-216(1987)
3.32 H. S. Jing, 'On the limitation principles for partial hybrid stress model', Computers & Structures, vol. 38, no. 1 113-117(1991).
3.33 T.H.H. Pian, 'Derivation of element stiffness matrices by assumed stress distribution', AIAA J., vol.2, no. 7, 1333-1336(1964).
3.34 E.F. Punch and S.N. Atluri, ' Development and testing of stable, invariant, isoparametric curvilinear 2- and 3-D hybrid- stress elements', Computer Methods in Applied Mechanics and Engineering, vol.47, 331-356(1984).

3.35 R. Rubinstein, E.F. Punch and S.N. Atluri, 'An analysis of, and remedies for, kinematic models in hybrid-stress finite elements: selection of stable, invariant stress fields', Computer Methods in Applied Mechanics and Engineering, vol. 38, 63-92(1983).

3.36 B.M. Fraeijs de Veubeke, 'Bending and stretching of plates -- special models for upper and lower bounds' Proceeding of the Conference on Matrix Methods in Structure Mechanics, AFFDL-TR-66-80, 863-886(1965).

3.37 T.H.H. Pian and P. Tong, ' Basis of finite element methods for solid continua', International Journal for Numerical Methods in Engineering, vol. 1, 3-28(1969).

3.38 R.D. Henshell, 'On hybrid finite elements', Proceeding of the Brunel University Conference of the Institute of Math and Its Applications, April, 1972.

Chapter 4

PARTIAL HYBRID ELEMENTS FOR ANALYSIS OF COMPOSITE LAMINATES

4.1 INTRODUCTION

A composite structure is usually made of hundreds of orthotropic laminae with different fibre orientations. The finite element analysis for composite structures is more difficult than that for structures made of isotropic materials. Due to the complex nature of composites, there are many different approaches to model them. In general, the finite elements for analysis of composites can be classified into three classes: 3-D solid elements, laminated elements, and multilayer elements [4.1]. They are formulated using two classes of composite structure models[4.2-4.3] as follows:

1. 3-D continuum models[4.4-4.6], in which each of the individual layers of a composite structure is treated as a three-dimensional continuum. Due to simplicity and efficiency, a special 3-D model, layer-wise models [4.13-4.16], is often used, in which displacement models are based on piecewise approximations of the response quantities in the thickness direction.

2. Equivalent single-layer plate/shell models[4.7-4.12], in which deformable models are based on global through-the-thickness displacement, strain and stress approximations;

In 3-D solid elements based on 3-D continuum models[4.4-4.6], no specific kinematic assumptions are introduced regarding the behaviour of a laminate. It takes the behaviour of the individual laminae into consideration. Therefore, the 3-D solid elements are used to accurately determine stresses in composite structures near discontinuities. However each layer in the laminate needs at least one element along the thickness of the structure. The number of unknowns in a finite element model will depend on the number of layers. Near the free edge of composite laminates, three or more elements along the thickness will be needed within a layer in order to accurately determine the transverse stresses with large gradient near interlaminar surfaces. In addition, 3-D solid elements show numerical instability under bending deformation when the aspect ratio is large. The aspect ratio is the ratio between the in-plane dimension and thickness dimension of the element. Usually, the thickness of a layer in a composite laminate is very small. So a fine finite element mesh in the in-plane dimensions is necessary because it not only needs to measure interlaminar stresses with large gradient, but also needs to maintain low aspect ratios. Thus, a full 3-D finite element modelling is computationally expensive and will quickly exhaust the computer space capacity.

In the laminated elements based on equivalent single-layer 2-D models[4.7-4.12], the variation in fiber orientations and material properties across the thickness is integrated to obtain a single property across the thickness. Therefore, in the finite element models, the number of unknowns through the thickness of a structure is independent of the number of layers in the composite. The laminated elements can be used to model the overall behaviour of composite structures reasonably well in problems such as vibration or buckling, but these may not provide useful results if interlaminar stresses are required.

In the multilayer elements based on layer-wise models [4.13-4.16], the 3-D discretization of a composite structure is separated into 2-D (in-plane) discretization and 1-D (thickness) discretization. Thus, the individual laminae are taken as 2-D layers or modeled by 3-D sub-elements. These layers and sub-elements are then assembled through the thickness. The layerwise models have the advantages over the previous models in that the data structure is 2-D and the number of degrees of freedom is less than that of a 3-D model. But, in these finite element models, the number of degrees of freedom is dependent on the number of layers in composite structures. A typical composite structure may have many layers, each of which requires one 2-D layer or one sub-element through the thickness. The number of degrees of freedom in the element is directly proportional to the number of layers in a laminate. Therefore, the number of unknowns in a finite element model is very large for laminates with many layers.

Finite elements can be classified not only in view of the composite structure models, but also in view of the assumption of the displacement and stress fields. In view of the assumption of displacement and stress within elements along the thickness of composite structures, finite elements can be divided into two categories: single-layer elements and multi-layer elements [4.17-4.18].

The single-layer element assumes a displacement field and/or a stress field over the element along the thickness direction. The number of displacement degrees of freedom in the element is independent of the number of material layers within the element. If the element contains only one material layer, it is a 3-D solid element; if the element contains more than one material layer along the thickness direction, the equivalent single-layer two-dimensional model must be used to obtain single properties across the thickness of the element, and it becomes a laminated element.

On the other hand, the multi-layer element assumes many displacement fields and/or stress fields within the element. Each displacement/stress field is related to a layer along the thickness of a composite laminate. The element matrices are assembled through the thickness by means of continuity conditions at the interfaces between different layers. The number of displacement degrees of freedom depends on the number of material layers in composite structures.

In this chapter, a series of partial hybrid finite elements will be developed using the composite variational principle. The general formulation of the partial hybrid element was given in chapter 3.

4.2 SINGLE-LAYER FINITE ELEMENTS

The single-layer finite elements include 3-D solids elements and laminated elements. Their element matrices can be formulated by means of the composite variational principle and expressed in a general form.

4.2.1 Fomulation of Partial Hybrid Single-Layer Element

The composite variation principle has been presented in the section 3.2, chapter 3. The variational functional is

$$\Pi_{co}=\int_V [E(\boldsymbol{q}) + \boldsymbol{\sigma}_g^T \boldsymbol{D}_L \boldsymbol{u} - \boldsymbol{F}^T \boldsymbol{u}]\, dV - \int_{S_t} \boldsymbol{T}^T \boldsymbol{u}\, dS \tag{4-1}$$

in which, the composite energy is

$$E(\boldsymbol{q}) = \frac{1}{2} \boldsymbol{q}^T [R]\, \boldsymbol{q} \tag{4-2}$$

and the vector of global variables includes in-plane strains and transverse stresses,

$$\boldsymbol{q} = \begin{Bmatrix} \boldsymbol{\varepsilon}_g \\ \boldsymbol{\sigma}_g \end{Bmatrix} \tag{4-3}$$

and the layer material matrix [R] is

$$[R]=\begin{bmatrix} R_1 & R_2 \\ R_2^T & R_3 \end{bmatrix}=\begin{bmatrix} S_1^{-1} & -S_1^{-1}S_2 \\ -S_2^T S_1^{-1} & S_2^T S_1^{-1} S_2 - S_3 \end{bmatrix} \tag{4-4}$$

or

$$[R]=\begin{bmatrix} R_1 & R_2 \\ R_2^T & R_3 \end{bmatrix}=\begin{bmatrix} C_1 - C_2 C_3^{-1} C_2^T & C_2 C_3^{-1} \\ C_3^{-1} C_2^T & -C_3^{-1} \end{bmatrix} \tag{4-4)'$$

where [S] is the compliance matrix of layer materials and [C] is the stiffness matrix of layer materials. Substituting equations (4-2)-(4-4) into equation (4-1), the functional becomes

$$\Pi_{co}=\int_V [\frac{1}{2}\boldsymbol{\varepsilon}_g^T [R_1]\,\boldsymbol{\varepsilon}_g+\frac{1}{2}\boldsymbol{\sigma}_g^T [R_3]\,\boldsymbol{\sigma}_g+\boldsymbol{\sigma}_g^T [R_2]^T \boldsymbol{\varepsilon}_g + \boldsymbol{\sigma}_g^T D_L \boldsymbol{u} - \boldsymbol{F}^T \boldsymbol{u}]\, dV - \int_{S_t} \boldsymbol{T}^T \boldsymbol{u}\, dS \tag{4-5}$$

Within a single-layer finite element (see figure 14), a displacement field is assumed along the thickness of the element. It is usually described by the nodal displacement δ,

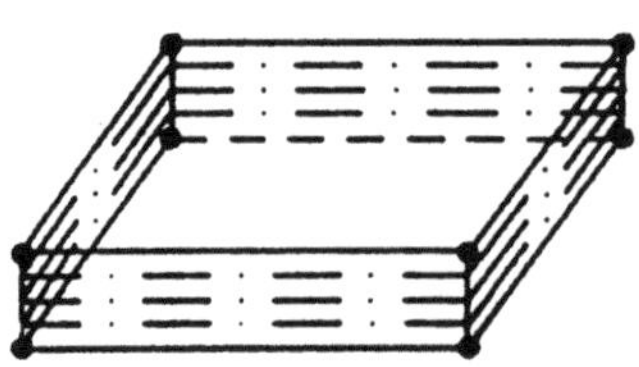

Figure 14 A Single-layer element

$$\boldsymbol{u}=\begin{Bmatrix} u \\ v \\ w \end{Bmatrix} = [N]\,\boldsymbol{\delta} \tag{4-6}$$

where [N] is the matrix of shape functions. Thus, the partial strains are

$$\boldsymbol{\varepsilon}_g=\begin{Bmatrix} \varepsilon_x \\ \varepsilon_y \\ \varepsilon_{xy} \end{Bmatrix} = \boldsymbol{D}_g\boldsymbol{u} = \left[\frac{\partial u}{\partial x}, \frac{\partial v}{\partial y}, \frac{\partial u}{\partial y}+\frac{\partial v}{\partial x}\right]^T = [B_g]\,\boldsymbol{\delta} \tag{4-7}$$

and the partial derivatives are

$$\boldsymbol{D}_L\boldsymbol{u} = \left[\frac{\partial w}{\partial z}, \frac{\partial w}{\partial y}+\frac{\partial v}{\partial z}, \frac{\partial w}{\partial x}+\frac{\partial u}{\partial z}\right]^T = [B_L]\,\boldsymbol{\delta} \tag{4-8}$$

in which, $[B_g]$ is a partial geometry matrix and $[B_L]$ is a partial derivative matrix. Along the thickness of composites, a partial stress field is also assumed independently as

$$\boldsymbol{\sigma}_g=\begin{Bmatrix} \sigma_z \\ \sigma_{yz} \\ \sigma_{zx} \end{Bmatrix} = [P_g]\,\boldsymbol{\beta} = [\boldsymbol{\sigma}_{g1}\boldsymbol{\sigma}_{g2}\cdots\boldsymbol{\sigma}_{gl}]\begin{Bmatrix} \beta_1 \\ \beta_2 \\ \cdot \\ \cdot \\ \beta_l \end{Bmatrix} \tag{4-9}$$

where $[P_g]$ is an assumed stress matrix, σ_{gj} are the partial stress modes, and β_j are the corresponding stress parameters. If the composite structure consists of N material layers, substituting equations (4-6)-(4-9) into the composite energy functional (4-5), the functional becomes

$$\begin{aligned}\Pi_{co}=\sum_{i=1}^{N}\{&\frac{1}{2}\boldsymbol{\delta}^{T}\int_{V}[B_g]^{T}[R_1^i][B_g]\,dV\,\boldsymbol{\delta}\\&+\frac{1}{2}\boldsymbol{\beta}^{T}\int_{V}[P_g]^{T}[R_3^i][P_g]\,dV\,\boldsymbol{\beta}\\&+\boldsymbol{\beta}^{T}\int_{V}[P_g]^{T}([B_L]+[R_2^i]^{T}[B_g])\,dV\,\boldsymbol{\delta}\}\\&-\boldsymbol{\delta}^{T}\int_{V}[N]^{T}\boldsymbol{F}^{i}dV\;-\boldsymbol{\delta}^{T}\int_{S}[N]^{T}\boldsymbol{T}^{i}dS\end{aligned} \tag{4-10}$$

Denote

$$\begin{aligned}[H]&=-\sum_{i=1}^{N}\int_V[P_g]^{T}[R_3^i][P_g]\,dV\\ [G]&=\sum_{i=1}^{N}\int_V[P_g]^{T}([B_L]+[R_2^i]^{T}[B_g])\,dV\\ [K_d]&=\sum_{i=1}^{N}\int_V[B_g]^{T}[R_1^i][B_g]\,dV\\ \boldsymbol{f}&=\sum_{i=1}^{N}\int_{V}[N]^{T}\boldsymbol{F}^{i}dV+\int_{S}[N]^{T}\boldsymbol{T}^{i}dS\end{aligned} \tag{4-11}$$

Note that, when the number of material layers is more than one, the variation in fibre orientations and material properties across the thickness of the element is integrated to obtain a single property across the thickness. Therefore, the size of the element matrices does not depend on the number of material layers in the element. Then, the functional can be expressed as

$$\Pi_{co}=\frac{1}{2}\boldsymbol{\delta}^{T}[K_d]\,\boldsymbol{\delta}-\frac{1}{2}\boldsymbol{\beta}^{T}[H]\,\boldsymbol{\beta}+\boldsymbol{\beta}^{T}[G]\,\boldsymbol{\delta}-\boldsymbol{\delta}^{T}\boldsymbol{f} \tag{4-12}$$

In this variational functional, there are two independent variables subject to variation. From the two partial stationary conditions with respect to β and δ as follows,

$$\frac{\partial \Pi_{co}}{\partial \boldsymbol{\beta}} = 0 \qquad and \qquad \frac{\partial \Pi_{co}}{\partial \boldsymbol{\delta}} = 0 \tag{4-13}$$

the relation between stress parameters β and nodal displacements δ is obtained,

$$[H]\,\boldsymbol{\beta} = [G]\,\boldsymbol{\delta} \tag{4-14}$$

and

$$[K_d]\,\boldsymbol{\delta} + [G]^T\boldsymbol{\beta} = \boldsymbol{f} \tag{4-15}$$

Eliminating β in the equation (4-14) and (4-15), one obtains

$$(\,[K_d] + [G]^T[H]^{-1}[G]\,)\,\boldsymbol{\delta} = \boldsymbol{f} \tag{4-16}$$

Denote

$$\begin{aligned} [K_h] &= [G]^T[H]^{-1}[G] \\ [K] &= [K_d] + [K_h] \end{aligned} \tag{4-17}$$

in which, the semi-stiffness matrix $[K_d]$ is a displacement-formulated stiffness matrix based on the globally continuous strains, and the semi-stiffness matrix $[K_h]$ is a hybrid-formulated stiffness matrix based on the globally continuous stresses. Then, the governing equation of the element is obtained,

$$[K]\,\boldsymbol{\delta} = \boldsymbol{f} \tag{4-18}$$

where [K] is the element stiffness matrix. For the partial hybrid element, the element stiffness matrix consists of a displacement-formulated stiffness matrix $[K_d]$ and a hybrid-formulated stiffness matrix $[K_h]$. In the single-layer element, the size of the element matrix [K] is not related to the number of material layers within the element. If there are more then one material layers, the single-layer element is a laminated element; if there is only one layer in the element, the element becomes a

3-D solid element.

After obtaining the nodal displacement δ of the elements in a finite element model, the displacement field, stress field, and strain field can be obtained using the following equations:

1. Displacement field

$$\mathbf{u}=\begin{Bmatrix} U \\ V \\ W \end{Bmatrix}=[N]\,\boldsymbol{\delta} \tag{4-6}$$

2. Partial globally continuous strains

$$\boldsymbol{\varepsilon}_g=\begin{Bmatrix} \varepsilon_x \\ \varepsilon_y \\ \varepsilon_{xy} \end{Bmatrix}=\mathbf{D}_g\mathbf{u}=[B_g]\,\boldsymbol{\delta} \tag{4-7}$$

3. Partial globally continuous stresses

$$\boldsymbol{\beta}=[H]^{-1}[G]\,\boldsymbol{\delta} \tag{4-14}$$

$$\boldsymbol{\sigma}_g=\begin{Bmatrix} \sigma_z \\ \sigma_{yz} \\ \sigma_{zx} \end{Bmatrix}=[P_g]\,\boldsymbol{\beta}=[P_g]\,[H]^{-1}[G]\,\boldsymbol{\delta} \tag{4-9}$$

4. Partial locally continuous stresses within i-th layer

$$\begin{aligned}\boldsymbol{\sigma}_L^i=\begin{Bmatrix} \sigma_x \\ \sigma_y \\ \sigma_{xy} \end{Bmatrix}&=[R_1^i]\,\boldsymbol{\varepsilon}_g+[R_2^i]\,\boldsymbol{\sigma}_g \\ &=\{[R_1^i]\,[B_g]+[R_2^i]\,[P_g]\,[H]^{-1}[G]\}\boldsymbol{\delta} \\ &=\{[S_1^i]^{-1}[B_g]+[C_2^i]\,[C_3^i]^{-1}[P_g]\,[H]^{-1}[G]\}\boldsymbol{\delta}\end{aligned} \tag{4-19}$$

5. Partial locally continuous strains within i-th layer

$$\boldsymbol{\varepsilon}_L^i=\begin{Bmatrix}\varepsilon_z\\ \varepsilon_{yz}\\ \varepsilon_{zx}\end{Bmatrix}=-[R_2^i]^T\boldsymbol{\varepsilon}_g-[R_3^i]\boldsymbol{\sigma}_g$$

$$=\{-[R_2^i]^T[B_g]-[R_3^i][P_g][H]^{-1}[G]\}\boldsymbol{\delta} \tag{4-20}$$

$$=\{[S_2^i]^T[S_1^i]^{-1}[B_g]+[C_3^i]^{-1}[P_g][H]^{-1}[G]\}\boldsymbol{\delta}$$

For convenience, all element matrices are given here again,

$$[K]=[K_d]+[K_h]$$

$$[K_d]=\sum_{i=1}^{N}\int_V[B_g]^T[R_1^i][B_g]\,dV$$

$$[K_h]=[G]^T[H]^{-1}[G] \tag{4-11}$$

$$[H]=-\sum_{i=1}^{N}\int_V[P_g]^T[R_3^i][P_g]\,dV$$

and (4-17)

$$[G]=\sum_{i=1}^{N}\int_V[P_g]^T([B_L]+[R_2^i]^T[B_g])\,dV$$

$$\boldsymbol{f}=\sum_{i=1}^{N}\int_V[N]^T\boldsymbol{F}^i dV+\int_S[N]^T\boldsymbol{T}^i dS$$

4.2.2 3-D Partial Hybrid Solid Element

In the element formulation above, when N=1, the single-layer element becomes a 3-D solid element because the element only contains a material layer. Many 3-D solid elements can be derived using the general formulation above. In this section, 3-D, 8-node and 20-node solid elements are presented.

a) 3-D, 8-node Partial Hybrid Solid Element

The 3-D, 8-node element (shown in figure 15) is a simplest finite element for 3-D analysis of structures. Therefore, it is firstly presented to show the formulation procedure of partial hybrid finite elements [4.19-4.21].

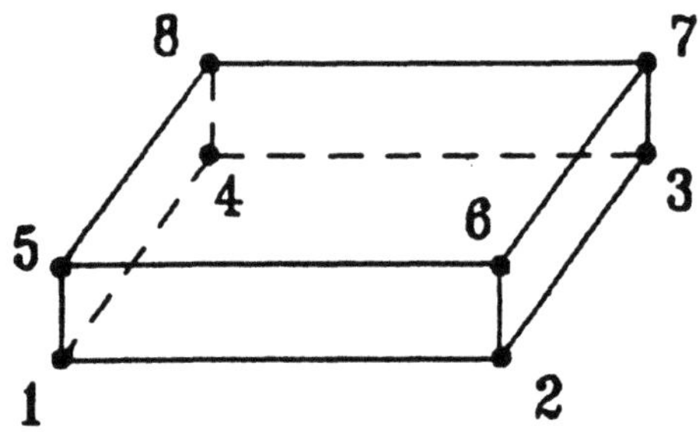

Figure 15 3-D, 8-node partial hybrid element

Geometry of Element

To map the element geometric shape, the global co-ordinates (x,y,z) of any point within the element can be written to interpolate the local co-ordinates (ξ, η, ζ) as follow:

$$x=\sum_{i=1}^{8} N_i x_i \qquad y=\sum_{i=1}^{8} N_i y_i \qquad z=\sum_{i=1}^{8} N_i z_i \tag{4-21}$$

where (x_i y_i z_i) are the global co-ordinates of the i-th node (i=1,2,...,8), and N_i are the shape functions which can be expressed as follows:

$$N_i=\frac{1}{8}(1+\xi_0)(1+\eta_0)(1+\zeta_0) \tag{4-22}$$

in which,

$$\xi_0=\xi_i\xi \qquad \eta_0=\eta_i\eta \qquad \zeta_0=\zeta_i\zeta \tag{4-23}$$

where ξ_i, η_i and ζ_i are the local co-ordinates of node i in the element parametric space.

Displacement Field

Within the element, a displacement field is assumed independently as follows:

$$u=\sum_{i=1}^{8} N_i u_i \qquad v=\sum_{i=1}^{8} N_i v_i \qquad w=\sum_{i=1}^{8} N_i w_i \tag{4-24}$$

where (u_i v_i w_i) are the i-th nodal displacements in the global co-ordinates system (i=1,2,...,8), and N_i are still the shape functions which are the same as that in the geometry formulation of the element (4-22)-(4-23). In the matrix form, the displacement field can be expressed as

$$\mathbf{u}=[N]\,\boldsymbol{\delta}=[N_1\mathbf{I}\ N_2\mathbf{I}\ \ldots\ N_8\mathbf{I}]\begin{Bmatrix}\boldsymbol{\delta}_1\\ \boldsymbol{\delta}_2\\ \cdots\\ \boldsymbol{\delta}_8\end{Bmatrix} \tag{4-25}$$

in which, [I] is a 3×3 unit matrix and the nodal displacement vector is

$$\boldsymbol{\delta}_i=\begin{Bmatrix}u_i\\ v_i\\ w_i\end{Bmatrix} \qquad i=1,2,\ldots,8 \tag{4-26}$$

Partial Strain Field and Partial Derivatives of the Displacement Field

Within the partial hybrid element, a partial strain field can be derived directly from the displacement field. It is

$$\boldsymbol{\varepsilon}_g=\begin{Bmatrix}\varepsilon_x\\ \varepsilon_y\\ \varepsilon_{xy}\end{Bmatrix}=\boldsymbol{D}_g\boldsymbol{u}=\begin{Bmatrix}\dfrac{\partial u}{\partial x}\\ \dfrac{\partial v}{\partial y}\\ \dfrac{\partial u}{\partial y}+\dfrac{\partial v}{\partial x}\end{Bmatrix}=[B_g]\boldsymbol{\delta} \tag{4-27}$$

and

$$[B_g]=[B_{g1}\ B_{g2}\ \cdots\ B_{g8}] \tag{4-28}$$

and

$$[B_{gi}]=\begin{bmatrix}N_{i,x} & 0 & 0\\ 0 & N_{i,y} & 0\\ N_{i,y} & N_{i,x} & 0\end{bmatrix} \tag{4-29}$$

Due to the fact that the partial strain-displacement relation (3-9) is satisfied a posterior, the locally continuous strains can not be derived directly and will be calculated using equation (4-20) after the nodal displacements having been obtained. But the partial derivatives of displacement field can be obtained as follows,

$$\boldsymbol{D}_L\boldsymbol{u}=\begin{Bmatrix}\dfrac{\partial w}{\partial z}\\ \dfrac{\partial v}{\partial z}+\dfrac{\partial w}{\partial y}\\ \dfrac{\partial w}{\partial x}+\dfrac{\partial u}{\partial z}\end{Bmatrix}=[B_L]\boldsymbol{\delta} \tag{4-30}$$

in which,

$$[B_L]=[B_{L1}\ B_{L2}\ \cdots\ B_{L8}] \tag{4-31}$$

and

$$[B_{Li}]=\begin{bmatrix} 0 & 0 & N_{i,z} \\ 0 & N_{i,z} & N_{i,y} \\ N_{i,z} & 0 & N_{i,x} \end{bmatrix} \tag{4-32}$$

where i=1,2,....,8. To map the derivatives from global co-ordinate system to local co-ordinate system, one can write

$$\begin{pmatrix} N_{i,\xi} \\ N_{i,\eta} \\ N_{i,\zeta} \end{pmatrix}=\begin{bmatrix} x_{,\xi} & y_{,\xi} & z_{,\xi} \\ x_{,\eta} & y_{,\eta} & z_{,\eta} \\ x_{,\zeta} & y_{,\zeta} & z_{,\zeta} \end{bmatrix}\begin{pmatrix} N_{i,x} \\ N_{i,y} \\ N_{i,z} \end{pmatrix}=[J]\begin{pmatrix} N_{i,x} \\ N_{i,y} \\ N_{i,z} \end{pmatrix} \tag{4-33}$$

where

$$x_{,\xi}=\sum_{i=1}^{8} N_{i,\xi}x_i, \quad \ldots \quad z_{,\zeta}=\sum_{i=1}^{8} N_{i,\zeta}z_i \tag{4-34}$$

The equation (4-33) can be rewritten

$$\begin{pmatrix} N_{i,x} \\ N_{i,y} \\ N_{i,z} \end{pmatrix}=[J]^{-1}\begin{pmatrix} N_{i,\xi} \\ N_{i,\eta} \\ N_{i,\zeta} \end{pmatrix} \tag{4-35}$$

in which

$$\begin{aligned} N_{i,\xi}&=\frac{1}{8}\xi_i(1+\eta_0)(1+\zeta_0) \\ N_{i,\eta}&=\frac{1}{8}\eta_i(1+\xi_0)(1+\zeta_0) \\ N_{i,\zeta}&=\frac{1}{8}\zeta_i(1+\xi_0)(1+\eta_0) \end{aligned} \tag{4-36}$$

For mapping the derivatives, it is convenient to introduce a radius vector and its derivatives:

$$\boldsymbol{r}=\begin{pmatrix} x \\ y \\ z \end{pmatrix} \qquad \boldsymbol{S}=\boldsymbol{r}_{,\xi}=\begin{pmatrix} x_{,\xi} \\ y_{,\xi} \\ z_{,\xi} \end{pmatrix} \tag{4-37}$$

$$\boldsymbol{T}=\boldsymbol{r}_{,\eta}=\begin{pmatrix} x_{,\eta} \\ y_{,\eta} \\ z_{,\eta} \end{pmatrix} \qquad \boldsymbol{V}=\boldsymbol{r}_{,\zeta}=\begin{pmatrix} x_{,\zeta} \\ y_{,\zeta} \\ z_{,\zeta} \end{pmatrix} \tag{4-38}$$

Then

$$[J]=[\boldsymbol{S}\ \boldsymbol{T}\ \boldsymbol{V}]^T \tag{4-39}$$

and

$$|J|=\boldsymbol{S}\cdot\boldsymbol{T}\times\boldsymbol{V} \tag{4-40}$$

and

$$[J]^{-1}=[\boldsymbol{T}\times\boldsymbol{V}\ \ \boldsymbol{V}\times\boldsymbol{S}\ \ \boldsymbol{S}\times\boldsymbol{T}]/|J| \tag{4-41}$$

Partial Stress Field

Within the element, a partial stress field is also assumed independently as follows,

$$\boldsymbol{\sigma}_g=\begin{Bmatrix} \sigma_z \\ \sigma_{yz} \\ \sigma_{zx} \end{Bmatrix}=[P_g]\,\boldsymbol{\beta}=[\boldsymbol{\sigma}_{g1}\boldsymbol{\sigma}_{g2}\ldots\boldsymbol{\sigma}_{gl}]\begin{Bmatrix} \beta_1 \\ \beta_2 \\ \cdot \\ \cdot \\ \beta_l \end{Bmatrix} \tag{4-9}$$

Using iso-function method (see section 3.4.2 in the chapter 3), an iso-function partial stress matrix can be derived from the displacement field (4-24) as follows [4.20],

$$[P_g] = \begin{bmatrix} 1 & 0 & 0 & \xi & 0 & 0 & \eta & 0 & 0 & \zeta & 0 & 0 & \xi\eta & 0 & 0 & \eta\zeta & 0 & \xi\zeta & 0 \\ 0 & 1 & 0 & 0 & \xi & 0 & 0 & \eta & 0 & 0 & \zeta & 0 & 0 & \xi\eta & 0 & 0 & 0 & 0 & \xi\zeta \\ 0 & 0 & 1 & 0 & 0 & \xi & 0 & 0 & \eta & 0 & 0 & \zeta & 0 & 0 & \xi\eta & 0 & \eta\zeta & 0 & 0 \end{bmatrix} \qquad (4\text{-}42)$$

In this partial stress matrix, there are 19 stress modes and some of them are not necessary. The unnecessary stress modes can be deleted by means of the classification method (see section 3.4.3 in the chapter 3).

Examination of Partial Hybrid Element

A major disadvantage of the hybrid stress finite element is the presence of spurious kinematic deformation modes. Therefore, a new hybrid element has to be examined. The stiffness matrix of a partial hybrid element is in the form,

$$[\boldsymbol{K}] = [\boldsymbol{K_d}] + [\boldsymbol{K_h}] \qquad (4\text{-}17)$$

There are two parts: a displacement-formulated stiffness matrix and a hybrid-formulated stiffness matrix. In order to avoid any kinematic deformation modes, the number of the stress modes in the assumed partial stress matrix $[P_g]$ must satisfy the following necessary condition

$$n_h \geq n - r - n_d \qquad (4\text{-}43)$$

in which, n is the total degrees of freedom of the element, r is the number of rigid body motions, and n_d is the rank of the displacement-formulated stiffness matrix $[K_d]$. The limitation principle indicates that a partial hybrid element is equivalent to its displacement counterpart if its partial stress field contains all displacement-derived stress modes. Due to the fact that iso-function stress matrix contains all displacement-derived stress modes (see discussion in the section 3.4.2, chapter 3) and conventional displacement element never has any kinematic deformation modes, the partial hybrid element using the iso-function partial stress matrix (4-42) is equivalent to its displacement counterpart. Eigenvalue examination of the element shows that the partial hybrid element has the same eigenvalues as its displacement counterpart and does not have any kinematic deformation modes. Therefore, the sufficient condition to avoid spurious kinematic deformation modes can be that the stress modes in the partial stress matrix are the same as that in the iso-function partial stress matrix. But there are a great number of unnecessary stress modes in the iso-function stress matrix. So, the classification method of stress modes (see discussion in the section 3.4.3, chapter 3) has to be used to take unnecessary stress modes out. Thus, a necessary and sufficient condition for guaranteeing the absence of kinematic deformation modes at the element level can be expressed as follows,

$$n_h = n - r - n_d \tag{4-44}$$

In the 3-D, 8-node element, there are (n=) 24 degrees of freedom and (r=) 6 degrees of the rigid body motion because each node has three components of displacements. Thus, the element has (n-r=) 18 natural deformation modes. By means of eigenvalue examination of the element, the rank of the partial stiffness matrix $[K_d]$ can be determined. Because the partial stiffness matrix $[K_d]$ gives 10 non-zero eigenvalues, the rank of the matrix $[K_d]$ is $(n_d=)$ 10 and the matrix $[K_d]$ represents 10 natural deformation modes of the element. In order to avoid any kinematic deformation modes, another partial stiffness matrix $[K_h]$ must give 8 non-zero eigenvalues and represent 8 natural deformation modes. According to the equation (4-44), the number of necessary stress modes is equal to 8. For this element, a partial stress matrix determined by the eigenfunction method [4.19] is

$$[P_g] = \begin{bmatrix} 1 & 0 & 0 & \xi & \eta & 0 & 0 & \xi\eta \\ 0 & 1 & 0 & 0 & 0 & \xi & -\xi & 0 \\ 0 & 0 & 1 & 0 & 0 & \eta & \eta & 0 \end{bmatrix} \tag{4-45}$$

This partial stress matrix only contains minimum number of stress modes. The examination of element shows that there is not any kinematic deformation modes.

b) 3-D, 20-node Partial Hybrid Solid Element

A 3-D, 20-node element [4.21] is shown in figure 16. It also can be obtained using the general formulation of single-layer element.

Geometry of Element

Firstly, the global co-ordinates (x,y,z) of any point within the element can be expressed in the form as follows:

$$x=\sum_{i=1}^{20} N_i x_i \quad y=\sum_{i=1}^{20} N_i y_i \quad z=\sum_{i=1}^{20} N_i z_i \tag{4-46}$$

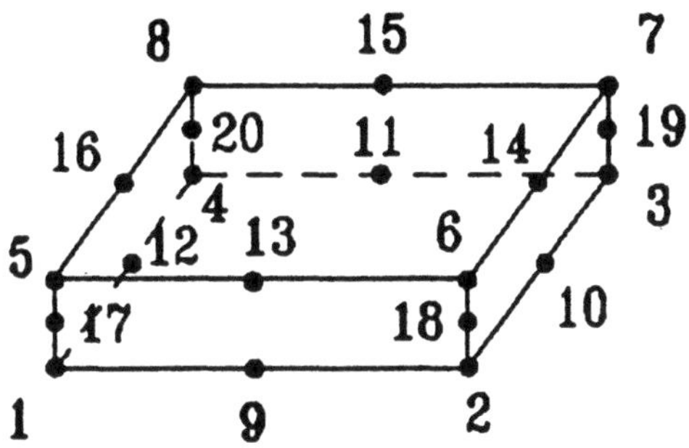

Figure 16 3-D, 20-node partial hybrid element

where (x_i y_i z_i) are the global co-ordinates of the i-th node (i=1,2,...,20), and N_i are the shape functions which are the functions of the local co-ordinates (ξ, η, ζ) as follows:

$$\begin{aligned} N_i = & \frac{1}{8}(1+\xi_0)(1+\eta_0)(1+\zeta_0)(\xi_0+\eta_0+\zeta_0-2)\xi_i^2\eta_i^2\zeta_i^2 \\ & +\frac{1}{4}(1-\xi^2)(1+\eta_0)(1+\zeta_0)(1-\xi_i^2)\eta_i^2\zeta_i^2 \\ & +\frac{1}{4}(1-\eta^2)(1+\zeta_0)(1+\xi_0)(1-\eta_i^2)\zeta_i^2\xi_i^2 \\ & +\frac{1}{4}(1-\zeta^2)(1+\xi_0)(1+\eta_0)(1-\zeta_i^2)\xi_i^2\eta_i^2 \end{aligned} \tag{4-47}$$

in which,

$$\xi_0=\xi_i\xi \quad \eta_0=\eta_i\eta \quad \zeta_0=\zeta_i\zeta \tag{4-48}$$

where ξ_i , η_i and ζ_i are the local co-ordinates of node i in the element parametric space.

Displacement Field

Within the element, a displacement field is assumed independently as follows:

$$u=\sum_{i=1}^{20} N_i u_i \qquad v=\sum_{i=1}^{20} N_i v_i \qquad w=\sum_{i=1}^{20} N_i w_i \tag{4-49}$$

where (u_i v_i w_i) are the i-th nodal displacements in the global co-ordinates system (i=1,2,...,20), and N_i are the shape functions (4-47)-(4-48). In the matrix form, the displacement field can be expressed as

$$\boldsymbol{u}=[N]\,\boldsymbol{\delta}=[N_1\boldsymbol{I}\ N_2\boldsymbol{I}\ \ldots\ N_{20}\boldsymbol{I}]\begin{Bmatrix}\boldsymbol{\delta}_1\\ \boldsymbol{\delta}_2\\ \ldots\\ \boldsymbol{\delta}_{20}\end{Bmatrix} \tag{4-50}$$

in which, [I] is a 3×3 unit matrix and the nodal displacement vector is

$$\boldsymbol{\delta}_i=\begin{Bmatrix}u_i\\ v_i\\ w_i\end{Bmatrix} \qquad i=1,2,\ldots,20 \tag{4-51}$$

Partial Strain Field and Partial Derivatives of the Displacement Field

Within the partial hybrid element, similar to the 8-node element above, a partial strain field can be derived directly from the displacement field (4-49). It is

$$\boldsymbol{\varepsilon}_g=\begin{Bmatrix}\varepsilon_x\\ \varepsilon_y\\ \varepsilon_{xy}\end{Bmatrix}=\boldsymbol{D}_g\boldsymbol{u}=\begin{Bmatrix}\dfrac{\partial u}{\partial x}\\ \dfrac{\partial v}{\partial y}\\ \dfrac{\partial u}{\partial y}+\dfrac{\partial v}{\partial x}\end{Bmatrix}=[B_g]\boldsymbol{\delta} \tag{4-52}$$

and

$$[B_g]=[B_{g1}\ B_{g2}\ \cdots\ B_{g20}] \tag{4-53}$$

and

$$[B_{gi}]=\begin{bmatrix} N_{i,x} & 0 & 0 \\ 0 & N_{i,y} & 0 \\ N_{i,y} & N_{i,x} & 0 \end{bmatrix} \tag{4-54}$$

Because the partial strain-displacement relation (3-9) is satisfied a posteriori, the partial derivatives of displacement field only can be derived from the displacement field as follows,

$$\boldsymbol{D}_L\boldsymbol{u}=\begin{Bmatrix} \dfrac{\partial w}{\partial z} \\ \dfrac{\partial v}{\partial z}+\dfrac{\partial w}{\partial y} \\ \dfrac{\partial w}{\partial x}+\dfrac{\partial u}{\partial z} \end{Bmatrix}=[B_L]\boldsymbol{\delta} \tag{4-55}$$

in which,

$$[B_L]=[B_{L1}\ B_{L2}\ \cdots\ B_{L20}] \tag{4-56}$$

and

$$[B_{Li}]=\begin{bmatrix} 0 & 0 & N_{i,z} \\ 0 & N_{i,z} & N_{i,y} \\ N_{i,z} & 0 & N_{i,x} \end{bmatrix} \tag{4-57}$$

where i=1,2,....,20. To map the derivatives from global co-ordinate system to local co-ordinate system, the following equations are used.

$$\begin{Bmatrix} N_{i,\xi} \\ N_{i,\eta} \\ N_{i,\zeta} \end{Bmatrix} = \begin{bmatrix} x_{,\xi} & y_{,\xi} & z_{,\xi} \\ x_{,\eta} & y_{,\eta} & z_{,\eta} \\ x_{,\zeta} & y_{,\zeta} & z_{,\zeta} \end{bmatrix} \begin{Bmatrix} N_{i,x} \\ N_{i,y} \\ N_{i,z} \end{Bmatrix} = [J] \begin{Bmatrix} N_{i,x} \\ N_{i,y} \\ N_{i,z} \end{Bmatrix} \tag{4-58}$$

where

$$x_{,\xi} = \sum_{i=1}^{20} N_{i,\xi} x_i, \quad \ldots \quad z_{,\zeta} = \sum_{i=1}^{20} N_{i,\zeta} z_i \tag{4-59}$$

The equation (4-58) can be rewritten as

$$\begin{Bmatrix} N_{i,x} \\ N_{i,y} \\ N_{i,z} \end{Bmatrix} = [J]^{-1} \begin{Bmatrix} N_{i,\xi} \\ N_{i,\eta} \\ N_{i,\zeta} \end{Bmatrix} \tag{4-60}$$

where

$$\begin{aligned} N_{i,\xi} = & \frac{1}{8}\xi_i(1+\eta_0)(1+\zeta_0)(2\xi_0+\eta_0+\zeta_0-1)\xi_i^2\eta_i^2\zeta_i^2 \\ & -\frac{1}{2}\xi(1+\eta_0)(1+\zeta_0)(1-\xi_i^2)\eta_i^2\zeta_i^2 \\ & +\frac{1}{4}\xi_i(1-\eta^2)(1+\zeta_0)(1-\eta_i^2)\zeta_i^2\xi_i^2 \\ & +\frac{1}{4}\xi_i(1-\zeta^2)(1+\eta_0)(1-\zeta_i^2)\xi_i^2\eta_i^2 \end{aligned} \tag{4-61}$$

$$\begin{aligned} N_{i,\eta} = & \frac{1}{8}\eta_i(1+\xi_0)(1+\zeta_0)(\xi_0+2\eta_0+\zeta_0-1)\xi_i^2\eta_i^2\zeta_i^2 \\ & +\frac{1}{4}\eta_i(1-\xi^2)(1+\zeta_0)(1-\xi_i^2)\eta_i^2\zeta_i^2 \\ & -\frac{1}{2}\eta(1+\zeta_0)(1+\xi_0)(1-\eta_i^2)\zeta_i^2\xi_i^2 \\ & +\frac{1}{4}\eta_i(1-\zeta^2)(1+\xi_0)(1-\zeta_i^2)\xi_i^2\eta_i^2 \end{aligned} \tag{4-62}$$

$$
\begin{aligned}
N_{i,\zeta}=&\frac{1}{8}\zeta_i(1+\xi_0)(1+\eta_0)(\xi_0+\eta_0+2\zeta_0-1)\xi_i^2\eta_i^2\zeta_i^2\\
&+\frac{1}{4}\zeta_i(1-\xi^2)(1+\eta_0)(1-\xi_i^2)\eta_i^2\zeta_i^2\\
&+\frac{1}{4}\zeta_i(1-\eta^2)(1+\xi_0)(1-\eta_i^2)\zeta_i^2\xi_i^2\\
&-\frac{1}{2}\zeta(1+\xi_0)(1+\eta_0)(1-\zeta_i^2)\xi_i^2\eta_i^2
\end{aligned}
\tag{4-63}
$$

For mapping the derivatives, a radius vector and its derivatives defined by equations (4-37)-(4-41) also can be used.

Partial Stress Field

Within the element, a partial stress field is assumed independently as follows,

$$
\boldsymbol{\sigma}_g=\begin{Bmatrix}\sigma_z\\ \sigma_{yz}\\ \sigma_{zx}\end{Bmatrix}=[P_g]\,\boldsymbol{\beta}=[\boldsymbol{\sigma}_{g1}\boldsymbol{\sigma}_{g2}\cdots\boldsymbol{\sigma}_{gl}]\begin{Bmatrix}\beta_1\\ \beta_2\\ \cdot\\ \cdot\\ \beta_l\end{Bmatrix}
\tag{4-9}
$$

Using iso-function method, an iso-function partial stress matrix can be derived from the displacement field (4-49) as follows [4.20],

$$
[P_g]=\left|\begin{array}{cccccccccccccccccc}
1&0&0&\xi&0&0&\eta&0&0&\zeta&0&0&\xi\eta&0&0&\xi\zeta&0&0\\
0&1&0&0&\xi&0&0&\eta&0&0&\zeta&0&0&\xi\eta&0&0&\xi\zeta&0\\
0&0&1&0&0&\xi&0&0&\eta&0&0&\zeta&0&0&\xi\eta&0&0&\xi\zeta
\end{array}\right.
$$

$$
\begin{array}{cccccccccccccccc}
\eta\zeta&0&0&\xi\eta\zeta&0&0&\xi^2&0&0&\eta^2&0&0&\zeta^2&0&0&\xi^2\eta\\
0&\eta\zeta&0&0&\xi\eta\zeta&0&0&\xi^2&0&0&\eta^2&0&0&\zeta^2&0&0\\
0&0&\eta\zeta&0&0&\xi\eta\zeta&0&0&\xi^2&0&0&\eta^2&0&0&\zeta^2&0
\end{array}
\tag{4-64}
$$

$$
\left.\begin{array}{ccccccccccccc}
0&0&\xi^2\zeta&0&\xi\eta^2&0&0&\eta^2\zeta&0&\xi\zeta^2&0&\eta\zeta^2&0\\
\xi^2\eta&0&0&\xi^2\zeta&0&\xi\eta^2&0&0&0&0&\xi\zeta^2&0&0\\
0&\xi^2\eta&0&0&0&0&\xi\eta^2&0&\eta^2\zeta&0&0&0&\eta\zeta^2
\end{array}\right|
$$

In this partial stress matrix, there are 47 stress modes. The examination of element will indicate that there are many unnecessary stress modes in this iso-function stress matrix.

Examination of Partial Hybrid Element

In the element, there are 20 nodes and each node has three components of displacements. Therefore, the element has (n=) 60 degrees of freedom. The degrees of the rigid displacement are equal to (r=) 6. Thus, the element has (n-r=) 54 natural deformation modes. The examination of the partial stiffness matrix $[K_d]$ gives 31 non-zero eigenvalues. So the rank of the partial stiffness matrix $[K_d]$ is (n_d=) 31 and it represents 31 natural deformation modes of the element. Another partial stiffness matrix $[K_h]$ must give 23 non-zero eigenvalues and represent 23 natural deformation modes in order to avoid any kinematic deformation modes mode. According to the equation (4-44), the assumed stress matrix must contain at least 23 stress modes.

In the iso-function stress matrix (4-64), there are 47 stress modes. It is more than double the number of necessary stress modes. The stress matrix (4-64) can be expressed in the form

$$[P_g] = [\sigma_1 \;\; \sigma_2 \;\; \sigma_3 \;\; \ldots \;\; \sigma_i \;\; \ldots \;\; \sigma_{47} \;] \tag{4-65}$$

Based on this iso-function stress matrix, the classification method gives an optimal stress matrix as follows,

$$[P_g] = \left| \begin{matrix} 1 & 0 & 0 & \xi & 0 & 0 & \eta & 0 & 0 & \zeta & 0 & 0 & \xi\eta & 0 & 0 & \xi\zeta & 0 & 0 \\ 0 & 1 & 0 & 0 & \xi & 0 & 0 & \eta & 0 & 0 & \zeta & 0 & 0 & \xi\eta & 0 & 0 & \xi\zeta & 0 \\ 0 & 0 & 1 & 0 & 0 & \xi & 0 & 0 & \eta & 0 & 0 & \zeta & 0 & 0 & \xi\eta & 0 & 0 & \xi\zeta \end{matrix} \right.$$

$$\left. \begin{matrix} \eta\zeta & 0 & \xi\eta\zeta & 0 & 0 \\ 0 & \eta\zeta & 0 & \xi\eta\zeta & 0 \\ 0 & 0 & 0 & 0 & \xi\eta\zeta \end{matrix} \right| \tag{4-66}$$

or

$$[P_g] = [\sigma_1 \;\; \sigma_2 \;\; \sigma_3 \;\; \ldots \;\; \sigma_{20} \;\; \sigma_{22} \;\; \ldots \;\; \sigma_{24} \;] \tag{4-66)'$$

Using this assumed partial stress matrix, two stiffness matrices of the 3-D, 20-node partial hybrid elements are examined. One is for isotropic material (see table 4); another is for anisotropic material (see table 5). In the tables,

$$\underline{\lambda_i} = \frac{\lambda_{hi}}{\lambda_{ui}} \tag{4-67}$$

where λ_{hi} is the eigenvalue of the partial hybrid element; λ_{ui} is the eigenvalue of the conventional displacement element. In tables 4 and 5, there are not any spurious zero eigenvalues. So the elements do not have any spurious kinematic deformation modes. From the results in tables 4 and 5, it can be concluded that if an assumed partial stress field can be used to construct a partial hybrid elements without kinematic deformation modes for the isotropic materials, it also can be used to construct the elements for anisotropic materials.

In order to study the effect of extra stress modes on the stiffness of elements, the assumed partial stress field consisted of the first 33 stress modes in iso-function stress matrix is examined. The results of the eigenvalue analysis are presented in table 6. Comparing the $\underline{\lambda_i}$ in the table 4 and in the table 6, it is shown that the eigenvalue $\underline{\lambda_i}$ of the element using 33 stress modes is larger than that using 23 stress modes. Therefore, the added stress modes stiffen the elements. One can examine a series of partial hybrid elements using different number of stress modes in assumed stress matrix. The examination will show that the more stress modes there are, the more stiffening the element is. If the number of added stress modes in the assumed stress matrix is increased sufficiently, the stiffness of the partial hybrid element will be equal to its conventional displacement counterpart. Such partial hybrid element has been presented by Han [4.20].

Table 4 Eigenvalue Analysis of Stiffness Matrix for the 3-D, 20-Node Element with 23 stress modes and isotropic materials: E=1100 GPa, ν=0.1

No.	λ_i	No.	λ_i	No.	λ_i
1	0.3822	19	0.9066	37	0.9594
2	0.3822	20	0.9106	38	0.9654
3	0.4759	21	0.9235	39	0.9694
4	0.5549	22	0.9272	40	0.9694
5	0.6403	23	0.9277	41	0.9694
6	0.6445	24	0.9296	42	0.9730
7	0.6846	25	0.9296	43	0.9782
8	0.7257	26	0.9366	44	0.9782
9	0.7257	27	0.9366	45	0.9782
10	0.7293	28	0.9376	46	0.9797
11	0.7679	29	0.9376	47	0.9881
12	0.7858	30	0.9389	48	0.9983
13	0.8467	31	0.9559	49	0.9983
14	0.8550	32	0.9559	50	0.9988
15	0.8657	33	0.9568	51	0.9989
16	0.8801	34	0.9573	52	1.0000
17	0.9001	35	0.9575	53	1.0000
18	0.9011	36	0.9575	54	1.0000

Table 5 Eigenvalue Analysis of Stiffness Matrix for the 3-D, 20-Node Element with 23 stress modes and anisotropic materials: E_L=174.6 GPa, E_T=7.0 GPa, G_{LT}=3.5 GPa, G_{TT}=1.4 GPa, $\nu_{12}=\nu_{13}=\nu_{23}$=0.25

No.	λ_i	No.	λ_i	No.	λ_i
1	0.5357	19	0.9641	37	0.9314
2	0.5888	20	0.9675	38	0.9994
3	0.5497	21	0.9206	39	0.9729
4	0.5013	22	0.9511	40	0.9593
5	0.6031	23	0.9316	41	0.9631
6	0.8726	24	0.8770	42	0.9807
7	0.7696	25	0.9044	43	0.9931
8	0.8002	26	0.9672	44	0.9442
9	0.9358	27	0.9759	45	1.0000
10	0.8494	28	0.9371	46	1.0000
11	0.5934	29	0.8017	47	0.9971
12	0.8294	30	0.8755	48	0.9975
13	0.8994	31	0.9618	49	0.9996
14	0.8053	32	0.9837	50	0.9961
15	0.8532	33	0.9952	51	0.9958
16	0.7674	34	0.9972	52	0.9975
17	0.7128	35	0.9974	53	0.9992
18	0.8763	36	0.9991	54	1.0000

Table 6 Eigenvalue Analysis of Stiffness Matrix for the 3-D, 20-Node Element with 33 stress modes and isotropic materials: E=1100 GPa, ν=0.1

No.	λ_i	No.	λ_i	No.	λ_i
1	0.4685	19	0.9971	37	1.0000
2	0.4685	20	0.9971	38	1.0000
3	0.7293	21	0.9981	39	1.0000
4	0.8803	22	0.9983	40	1.0000
5	0.9014	23	0.9983	41	1.0000
6	0.9014	24	0.9988	42	1.0000
7	0.9277	25	1.0000	43	1.0000
8	0.9296	26	1.0000	44	1.0000
9	0.9296	27	1.0000	45	1.0000
10	0.9389	28	1.0000	46	1.0000
11	0.9497	29	1.0000	47	1.0000
12	0.9556	30	1.0000	48	1.0000
13	0.9573	31	1.0000	49	1.0000
14	0.9575	32	1.0000	50	1.0000
15	0.9575	33	1.0000	51	1.0000
16	0.9730	34	1.0000	52	1.0000
17	0.9782	35	1.0000	53	1.0000
18	0.9878	36	1.0000	54	1.0000

4.2.3 Partial Hybrid Laminated Element

In the general formulation of single-layer element above, when N>1, the element will contain more than one material layer and the single-layer element will become a laminated element.

a) 3-D, 20-node Partial Hybrid Laminated Element

The formulation of a 3-D, 20-node laminated element (see figure 17) is the same as that of 3-D, 20-node partial hybrid solid element (4-46)-(4-66). But the variation in fibre orientations and material properties across the thickness of the element must be integrated in order to obtain a single property, and element matrices are calculated by equation (4-11). Using the assumed stress matrix (4-66), a laminated element with fibre orientation [90, 0, 90] is examined. The results of the examination of the element is given in table 7. The results show that the element does not have any kinematic deformation modes.

Figure 17 3-D, 20-node laminated element

Table 7 Eigenvalue Analysis of Stiffness Matrix for the 3-D, 20-Node Laminated Element [90, 0,90] with 23 stress modes and materials: E_L=174.6 GPa, E_T=7.0 GPa, G_{LT}=3.5 GPa, G_{TT}=1.4 GPa, $\nu_{12}=\nu_{13}=\nu_{23}$=0.25

No.	λ_i	No.	λ_i	No.	λ_i
1	0.4166	19	0.9312	37	0.9274
2	0.4103	20	0.9271	38	0.9993
3	0.4816	21	0.9425	39	0.9171
4	0.4492	22	0.9026	40	0.9443
5	0.6359	23	0.9213	41	0.9353
6	0.6520	24	0.8665	42	0.9221
7	0.6492	25	0.8663	43	0.9753
8	0.5950	26	0.9470	44	1.0000
9	0.8048	27	0.9141	45	0.9993
10	0.7801	28	0.8579	46	0.9930
11	0.7075	29	0.8483	47	0.9958
12	0.6489	30	0.9318	48	0.9974
13	0.9784	31	0.9251	49	0.9994
14	0.7856	32	0.9448	50	0.9995
15	0.8342	33	0.9778	51	0.9939
16	0.9216	34	0.9442	52	0.9991
17	0.8371	35	0.9889	53	1.0000
18	0.8981	36	1.0000	54	1.0000

b) 4-node Partial Hybrid Degenerated Plate Element

The degenerated plate element was originally introduced by Ahead, Irons and Zienkiewicz [4.22] for the linear analysis of moderately thick and thin shells. Chao and Reddy [4.23] presented a degenerated element based on the total Lagrangian description of the motion of a layered anisotropic composite medium. In chapter 1, a 4-node degenerated plate element has been presented using displacement element formulation. But similar to the plate/shell elements based on the 2-D plate/shell theories, for analysis of composites, the degenerated plate/shell elements using conventional displacement element formulation suffer from a common deficiency: constitutive equations lead to discontinuous interlaminar stresses. Equilibrium equations have been often used in recovering the interlaminar stresses. Regardless of the controversy and complexity, the use of equilibrium equations will reduce the accuracy of the stresses owing to the numerical differentiation. However, partial hybrid elements formulation can overcome the stress continuity limitations of single-layer models due to the fact that a partial stress field is assumed independently. Here, the 4-node degenerated plate element is presented again using partial hybrid element formulation. Its number of degrees of freedom per node is also independent from the number of layers in a composite structure.

<u>Geometry of Element</u>

Firstly, consider a typical thick plate element in figure 18. The co-ordinates of a typical point in the element can be written as

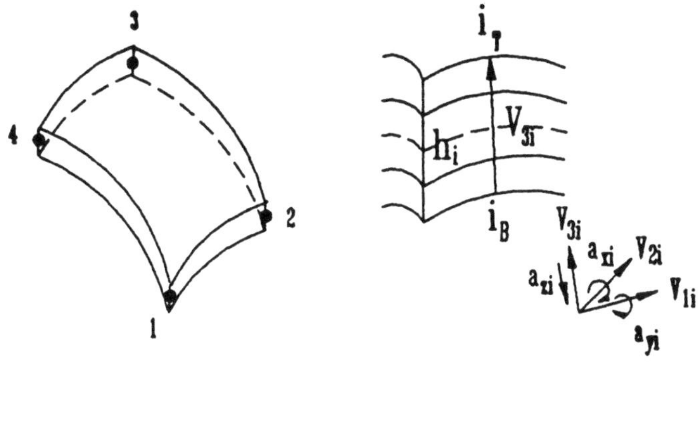

(a) (b)

Figure 18 A degenerated plate element

$$\begin{Bmatrix} x \\ y \\ z \end{Bmatrix} = \sum_{i=1}^{4} N_i(\xi,\eta)\frac{1+\zeta}{2}\begin{Bmatrix} x_i \\ y_i \\ z_i \end{Bmatrix}_{top} + \sum_{i=1}^{4} N_i(\xi,\eta)\frac{1-\zeta}{2}\begin{Bmatrix} x_i \\ y_i \\ z_i \end{Bmatrix}_{bottom} \tag{4-48}$$

where $N_i(\xi,\eta)$ are shape functions, ξ and η are the normalized curvilinear co-ordinates in the middle plane of the plate, ζ is a linear co-ordinate in the thickness direction and only approximately normal to the middle surface, and (x_i, y_i, z_i) are the global co-ordinates at node i. The shape functions are

$$N_i = \frac{1}{4}(1+\xi_0)(1+\eta_0) \tag{4-69}$$

in which,

$$\xi_0 = \xi_i \xi \quad \eta_0 = \eta_i \eta \quad (i = 1,2,3,4) \tag{4-70}$$

This equation can be rewritten in the form specified by the 'vector' connecting the upper and lower points (shown in figure 18) and the mid-surface co-ordinates as

$$\begin{Bmatrix} x \\ y \\ z \end{Bmatrix} = \sum_{i=1}^{4} N_i(\xi,\eta)\begin{Bmatrix} x_i \\ y_i \\ z_i \end{Bmatrix} + \sum_{i=1}^{4} N_i(\xi,\eta)\, h_i \frac{\zeta}{2} \mathbf{V}_{3i} \tag{4-71}$$

where

$$\mathbf{V}_{3i} = \begin{Bmatrix} l_{3i} \\ m_{3i} \\ n_{3i} \end{Bmatrix} = \frac{1}{h_i}\left(\begin{Bmatrix} x_i \\ y_i \\ z_i \end{Bmatrix}_T - \begin{Bmatrix} x_i \\ y_i \\ z_i \end{Bmatrix}_B\right) \tag{4-72}$$

and

$$h_i = \sqrt{(x_{iT}-x_{iB})^2 + (y_{iT}-y_{iB})^2 + (z_{iT}-z_{iB})^2} \tag{4-73}$$

Displacement Field

In the element, the displacement field is assumed as a continuous field through the entire thickness of a composite structure. Although there are numerous plate theories which include transverse shear deformations in the literature, the transverse normal stress is always not taken into account. Actually, the hypothesis $\varepsilon_z = 0$ (or an equivalent hypothesis) should not be used [4.24] in order to construct the consistent high-order theory. For analysis of composite, the first-order shear deformation theory has to be improved. For this element, the following displacement field is assumed [4.25-4.26],

$$\begin{aligned} u &= u_0 + z\, a_x \\ v &= v_0 + z\, a_y \\ w &= w_0 + z\, a_z \end{aligned} \tag{4 - 74}$$

In this displacement field, it is assumed that a line that is straight and normal to the middle surface before deformation is still straight, but not necessarily 'normal' to the middle surface after deformation. Thus, the displacement throughout the element will be uniquely defined by three Cartesian components (u_i, v_i and w_i) of the displacement at the mid-surface node i, two rotations (a_{xi} and a_{yi}) of the nodal vector $\mathbf{V}_{3i}$ about orthogonal directions normal to it, and one transverse normal deformation (a_{zi}) in the thickness direction. Based on this assumption, the i-th nodal displacement can be expressed as

$$\boldsymbol{u_i} = \begin{Bmatrix} u_i \\ v_i \\ w_i \end{Bmatrix} + \frac{h_i}{2}\zeta \,[\boldsymbol{V_{1i}} \;\; -\boldsymbol{V_{2i}} \;\; \mathbf{0}] \begin{Bmatrix} a_{xi} \\ a_{yi} \\ 0 \end{Bmatrix} + \frac{h_i}{2}\zeta \,[\mathbf{0} \;\; \mathbf{0} \;\; \boldsymbol{V_{3i}}] \begin{Bmatrix} 0 \\ 0 \\ a_{zi} \end{Bmatrix} \tag{4-75}$$

in which, $\mathbf{V}_{1i}$, $\mathbf{V}_{2i}$ and $\mathbf{V}_{3i}$ are the unit vectors of the local co-ordinate (ξ, η, ζ) at node i. They can be calculated as follows:

$$\boldsymbol{V_{1i}} = \begin{Bmatrix} l_{1i} \\ m_{1i} \\ n_{1i} \end{Bmatrix} = \frac{\boldsymbol{i} \times \boldsymbol{V_{3i}}}{|\boldsymbol{i} \times \boldsymbol{V_{3i}}|} \qquad \boldsymbol{V_{2i}} = \begin{Bmatrix} l_{2i} \\ m_{2i} \\ n_{2i} \end{Bmatrix} = \boldsymbol{V_{3i}} \times \boldsymbol{V_{1i}} \tag{4-76}$$

If $\mathbf{i} \times \mathbf{V}_{3i} = 0$, $\mathbf{i}$ can be replaced by $\mathbf{j}$. Thus, the displacement field is

$$\begin{Bmatrix} u \\ v \\ w \end{Bmatrix} = \sum_{i=1}^{4} N_i \left[\begin{Bmatrix} u_i \\ v_i \\ w_i \end{Bmatrix} + \frac{\zeta}{2} h_i \begin{bmatrix} l_{1i} & -l_{2i} & l_{3i} \\ m_{1i} & -m_{2i} & m_{3i} \\ n_{1i} & -n_{2i} & n_{3i} \end{bmatrix} \begin{Bmatrix} a_{xi} \\ a_{yi} \\ a_{zi} \end{Bmatrix} \right] \tag{4-77}$$

They can be rewritten in the form

$$\begin{Bmatrix} u \\ v \\ w \end{Bmatrix} = \sum_{i=1}^{4} N_i \left[\begin{Bmatrix} u_i \\ v_i \\ w_i \end{Bmatrix} + \zeta \, [b_i] \begin{Bmatrix} a_{xi} \\ a_{yi} \\ a_{zi} \end{Bmatrix} \right] = \sum_{i=1}^{4} [N]_i \boldsymbol{\delta}_i \tag{4-78}$$

where

$$[b_i] = \begin{bmatrix} b_{11i} & b_{12i} & b_{13i} \\ b_{21i} & b_{22i} & b_{23i} \\ b_{31i} & b_{32i} & b_{33i} \end{bmatrix} = \frac{h_i}{2} \, [\boldsymbol{V}_{1i} \; -\boldsymbol{V}_{2i} \; \boldsymbol{V}_{3i}] \tag{4-79}$$

$$\boldsymbol{\delta}_i = [u_i \; v_i \; w_i \; a_{xi} \; a_{yi} \; a_{zi}]^T \tag{4-80}$$

$$[N]_i = \begin{bmatrix} N_i & 0 & 0 & N_i\zeta b_{11i} & N_i\zeta b_{12i} & N_i\zeta b_{13i} \\ 0 & N_i & 0 & N_i\zeta b_{21i} & N_i\zeta b_{22i} & N_i\zeta b_{23i} \\ 0 & 0 & N_i & N_i\zeta b_{31i} & N_i\zeta b_{32i} & N_i\zeta b_{33i} \end{bmatrix} \tag{4-81}$$

Partial Strain Field and Partial Derivatives of the Displacement Field

The partial globally continuous strains can be derived from the displacement field as follows,

$$\boldsymbol{\varepsilon}_g = \begin{Bmatrix} \varepsilon_x \\ \varepsilon_y \\ \varepsilon_{xy} \end{Bmatrix} = \boldsymbol{D}_g \boldsymbol{u} = \begin{Bmatrix} \dfrac{\partial u}{\partial x} \\ \dfrac{\partial v}{\partial y} \\ \dfrac{\partial u}{\partial y} + \dfrac{\partial v}{\partial x} \end{Bmatrix} = [B_g] \boldsymbol{\delta} \tag{4-82}$$

in which,

$$[B_g] = [B_{g1} \; B_{g2} \; \cdots \; B_{g4}] \tag{4-83}$$

and

$$\boldsymbol{\delta} = [\boldsymbol{\delta}_1 \; \boldsymbol{\delta}_2 \; \boldsymbol{\delta}_3 \; \boldsymbol{\delta}_4]^T \tag{4-84}$$

The partial geometric matrix at the i-th node is,

$$[B_{gi}] = \begin{bmatrix} N_{i,x} & 0 & 0 & b_{11i}c_{ix} & b_{12i}c_{ix} & b_{13i}c_{ix} \\ 0 & N_{i,y} & 0 & b_{21i}c_{iy} & b_{22i}c_{iy} & b_{23i}c_{iy} \\ N_{i,y} & N_{i,x} & 0 & b_{11i}c_{iy}+b_{21i}c_{ix} & b_{12i}c_{iy}+b_{22i}c_{ix} & b_{13i}c_{iy}+b_{23i}c_{ix} \end{bmatrix} \tag{4-85}$$

where

$$\begin{aligned} c_{ix} &= N_{i,x}\zeta + N_i\zeta_{,x} \\ c_{iy} &= N_{i,y}\zeta + N_i\zeta_{,y} \\ c_{iz} &= N_{i,z}\zeta + N_i\zeta_{,z} \end{aligned} \tag{4-86}$$

The partial derivatives of displacement field can be also derived from the displacement field as follows,

$$\boldsymbol{D}_L\boldsymbol{u} = \begin{Bmatrix} \dfrac{\partial w}{\partial z} \\ \dfrac{\partial v}{\partial z} + \dfrac{\partial w}{\partial y} \\ \dfrac{\partial w}{\partial x} + \dfrac{\partial u}{\partial z} \end{Bmatrix} = [B_L]\boldsymbol{\delta} \tag{4-87}$$

in which,

$$[B_L] = [B_{L1} \; B_{L2} \; \ldots \; B_{L4}] \tag{4-88}$$

and

$$[B_{Li}] = \left| \begin{matrix} 0 & 0 & N_{i,z} & b_{31i}c_{iz} \\ 0 & N_{i,z} & N_{i,y} & b_{21i}c_{iz}+b_{31i}c_{iy} \\ N_{i,z} & 0 & N_{i,x} & b_{31i}c_{ix}+b_{11i}c_{iz} \end{matrix} \right. \qquad (4\text{-}89)$$

$$\left. \begin{matrix} b_{32i}c_{iz} & b_{33i}c_{iz} \\ b_{22i}c_{iz}+b_{32i}c_{iy} & b_{23i}c_{iz}+b_{33i}c_{iy} \\ b_{32i}c_{ix}+b_{12i}c_{iz} & b_{33i}c_{ix}+b_{13i}c_{iz} \end{matrix} \right|$$

In order to calculate $N_{i,x}$, $N_{i,y}$, $N_{i,z}$ and $\zeta_{,x}$, $\zeta_{,y}$, $\zeta_{,z}$, the following vectors are introduced:

$$\boldsymbol{S}=\begin{Bmatrix} x_{,\xi} \\ y_{,\xi} \\ z_{,\xi} \end{Bmatrix}=\sum_{i=1}^{4} N_{i,\xi}\left(\begin{Bmatrix} x_i \\ y_i \\ z_i \end{Bmatrix}+\frac{h_i}{2}\zeta \boldsymbol{V}_{3i}\right) \qquad (4\text{-}90)$$

$$\boldsymbol{T}=\begin{Bmatrix} x_{,\eta} \\ y_{,\eta} \\ z_{,\eta} \end{Bmatrix}=\sum_{i=1}^{4} N_{i,\eta}\left(\begin{Bmatrix} x_i \\ y_i \\ z_i \end{Bmatrix}+\frac{h_i}{2}\zeta \boldsymbol{V}_{3i}\right) \qquad (4\text{-}91)$$

$$\boldsymbol{V}=\begin{Bmatrix} x_{,\zeta} \\ y_{,\zeta} \\ z_{,\zeta} \end{Bmatrix}=\sum_{i=1}^{4} N_i \frac{h_i}{2} \boldsymbol{V}_{3i} \qquad (4\text{-}92)$$

then, the Jacobian matrix is

$$[\, J \,] = [\, \mathbf{S} \;\; \mathbf{T} \;\; \mathbf{V} \,]^T \qquad (4 - 93)$$

$$[\, J \,]^{-1} = [\, \mathbf{T}\times\mathbf{V} \;\; \mathbf{V}\times\mathbf{S} \;\; \mathbf{S}\times\mathbf{T} \,] \,/\, |\, J \,| \qquad (4 - 94)$$

$$|\, J \,| = \mathbf{S}\times\mathbf{T}\cdot\mathbf{V} \qquad (4 - 95)$$

Because

$$\begin{Bmatrix} N_{i,\xi} \\ N_{i,\eta} \\ N_{i,\zeta} \end{Bmatrix} = \begin{bmatrix} x_{,\xi} & y_{,\xi} & z_{,\xi} \\ x_{,\eta} & y_{,\eta} & z_{,\eta} \\ x_{,\zeta} & y_{,\zeta} & z_{,\zeta} \end{bmatrix} \begin{Bmatrix} N_{i,x} \\ N_{i,y} \\ N_{i,z} \end{Bmatrix} = [J] \begin{Bmatrix} N_{i,x} \\ N_{i,y} \\ N_{i,z} \end{Bmatrix} \tag{4-96}$$

the derivatives of shape function with respect to global co-ordinates are

$$\begin{Bmatrix} N_{i,x} \\ N_{i,y} \\ N_{i,z} \end{Bmatrix} = \begin{bmatrix} \xi_{,x} & \eta_{,x} & \zeta_{,x} \\ \xi_{,y} & \eta_{,y} & \zeta_{,y} \\ \xi_{,z} & \eta_{,z} & \zeta_{,z} \end{bmatrix} \begin{Bmatrix} N_{i,\xi} \\ N_{i,\eta} \\ N_{i,\zeta} \end{Bmatrix} = [J]^{-1} \begin{Bmatrix} N_{i,\xi} \\ N_{i,\eta} \\ N_{i,\zeta} \end{Bmatrix} \tag{4-97}$$

Due to $N_{i,\zeta} = 0$, the expression (4-97) can be rewritten as

$$\begin{Bmatrix} N_{i,x} \\ N_{i,y} \\ N_{i,z} \end{Bmatrix} = [\mathbf{T}\times\mathbf{V} \;\; \mathbf{V}\times\mathbf{S}] \,/|J| \begin{Bmatrix} N_{i,\xi} \\ N_{i,\eta} \end{Bmatrix} \tag{4-98}$$

and

$$\begin{Bmatrix} \zeta_{,x} \\ \zeta_{,y} \\ \zeta_{,z} \end{Bmatrix} = \frac{\mathbf{S}\times\mathbf{T}}{|J|} \tag{4-99}$$

The geometric matrix [B'] in the local co-ordinate system can be obtained by means of transformation matrix [T],

$$[B'] = \begin{Bmatrix} B'_g \\ B'_L \end{Bmatrix} = [T_B] \begin{Bmatrix} B_g \\ B_L \end{Bmatrix} \tag{4-100}$$

and

$$[T_B]=\begin{bmatrix} l_1^2 & m_1^2 & l_1m_1 & n_1^2 & m_1n_1 & n_1l_1 \\ l_2^2 & m_2^2 & l_2m_2 & n_2^2 & m_2n_2 & n_2l_2 \\ 2l_1l_2 & 2m_1m_2 & l_1m_2+l_2m_1 & 2n_1n_2 & m_1n_2+m_2n_1 & n_1l_2+n_2l_1 \\ l_3^2 & m_3^2 & l_3m_3 & n_3^2 & m_3n_3 & n_3l_3 \\ 2l_2l_3 & 2m_2m_3 & l_2m_3+l_3m_2 & 2n_2n_3 & m_2n_3+m_3n_2 & n_2l_3+n_3l_2 \\ 2l_3l_1 & 2m_3m_1 & l_3m_1+l_1m_3 & 2n_3n_1 & m_3n_1+m_1n_3 & n_3l_1+n_1l_3 \end{bmatrix} \tag{4-101}$$

$[T_B]$ is the transformation matrix for the derivatives of displacements from global co-ordinate (x, y, z) to local co-ordinate. The direction cosines of the local co-ordinates are

$$\mathbf{V_3}=\begin{Bmatrix} l_3 \\ m_3 \\ n_3 \end{Bmatrix}=\frac{\sum_{i=1}^{4} N_i \mathbf{V_{3i}}}{\left|\sum_{i=1}^{4} N_i \mathbf{V_{3i}}\right|} \tag{4-102}$$

and

$$\mathbf{V_1}=\begin{Bmatrix} l_1 \\ m_1 \\ n_1 \end{Bmatrix}=\frac{\mathbf{i}\times\mathbf{V_3}}{|\mathbf{i}\times\mathbf{V_3}|} \qquad \mathbf{V_2}=\begin{Bmatrix} l_2 \\ m_2 \\ n_2 \end{Bmatrix}=\mathbf{V_3}\times\mathbf{V_1} \tag{4-103}$$

Assumed Partial Stress Field

In the element, the partial stress field is assumed independently as continuous functions along the thickness of a composite structure.

$$\boldsymbol{\sigma_g}=\begin{Bmatrix} \sigma_z \\ \tau_{yz} \\ \tau_{zx} \end{Bmatrix}=[P_g]\,\boldsymbol{\beta}=[T]\,[P]\,\boldsymbol{\beta}=[T]\,[\boldsymbol{\sigma_1}\ \boldsymbol{\sigma_2}\ldots\boldsymbol{\sigma_m}]\begin{Bmatrix} \beta_1 \\ \beta_2 \\ \cdots \\ \beta_m \end{Bmatrix} \tag{4-104}$$

where, $\boldsymbol{\beta}$ is the stress parameter vector. The matrix [T] is a multiplying matrix and is determined by the traction conditions on the top and bottom surfaces of the structure. For example, if there is a distributed normal load on the upper surface, the transverse shear stresses must be equal to zero on both surfaces of top and bottom, and the transverse normal stress must be equal to zero on the bottom surface and be equal to the distributed load on the top surface. Therefore, the multiplying matrix has to be assumed as

$$[T] = \begin{bmatrix} 1+\zeta & 0 & 0 \\ 0 & 1-\zeta^2 & 0 \\ 0 & 0 & 1-\zeta^2 \end{bmatrix} \tag{4-105}$$

The matrix [P] consists of a group of stress modes which can be derived directly from the assumed displacement field using the iso-function method. The iso-function partial stress matrix of the element is

$$[P] = \begin{bmatrix} 1 & 0 & 0 & \xi & 0 & 0 & \eta & 0 & 0 & \zeta & 0 & 0 & \xi\eta & 0 & 0 & \eta\zeta & 0 & \xi\zeta & 0 \\ 0 & 1 & 0 & 0 & \xi & 0 & 0 & \eta & 0 & 0 & \zeta & 0 & 0 & \xi\eta & 0 & 0 & 0 & 0 & \xi\zeta \\ 0 & 0 & 1 & 0 & 0 & \xi & 0 & 0 & \eta & 0 & 0 & \zeta & 0 & 0 & \xi\eta & 0 & \eta\zeta & 0 & 0 \end{bmatrix} \tag{4-106}$$

Examination of Partial Hybrid Element

In the element, there are (n=) 24 degrees of freedom and (r=) 6 degrees of the rigid displacement. The element has (n-r=) 18 natural deformation modes. The eigenvalue examination of the stiffness matrix $[K_d]$ gives 10 non-zero eigenvalues. So the rank of the partial stiffness matrix $[K_d]$ is $(n_d=)$ 10. Thus, 10 natural deformation modes of the element can be represented by the partial stiffness matrix $[K_d]$, and $(n\text{-}r\text{-}n_d=)$ 8 natural deformation modes must be represented by another partial stiffness matrix $[K_h]$ in order to avoid any kinematic deformation modes. According to the equation (4-44), the assumed stress matrix must contain at least 8 stress modes.

Based on the iso-function partial stress matrix (4-106), when the multiplying matrix [T] is an unit matrix [I], the classification method gives an optimal stress matrix as follows,

$$[P] = \begin{bmatrix} 1 & 0 & 0 & \xi & 0 & \eta & 0 & \xi\eta \\ 0 & 1 & 0 & 0 & 0 & 0 & \eta & 0 \\ 0 & 0 & 1 & 0 & \xi & 0 & 0 & 0 \end{bmatrix} \tag{4-107}$$

Using this assumed partial stress matrix, the partial hybrid degenerated plate element is examined. The results of eigenvalue examination are given in Table 8. In the table, λ_i is the eigenvalue of the elements.

Table 8 Eigenvalue of Stiffness Matrix for 4-node Degenerated Hybrid Element with 8 stress modes and E=1100 GPa, ν=0.3

No.	λ_i (*10^3)	No.	λ_i (*10^3)	No.	λ_i (*10^3)
1	0.2821	7	0.6822	13	1.6920
2	0.2821	8	0.6822	14	1.6920
3	0.3291	9	1.0480	15	1.6920
4	0.3626	10	1.1280	16	1.6920
5	0.3626	11	1.5740	17	1.6920
6	0.5641	12	1.5740	18	5.5000

On the free-traction surface, the transverse stresses must be zero in order to satisfy the boundary condition. In this case, the multiplying matrix is not a unit matrix. For example, if free traction condition is applied on both top and bottom surfaces, the multiplying matrix is

$$[T_1] = \begin{bmatrix} 1-\zeta^2 & 0 & 0 \\ 0 & 1-\zeta^2 & 0 \\ 0 & 0 & 1-\zeta^2 \end{bmatrix} \tag{4-108}$$

The eigenvalues of the element are given in table 9. If the free traction condition is only applied on bottom surface, then the multiplying matrix becomes

$$[T_2] = \frac{1}{2}\begin{bmatrix} 1+\zeta & 0 & 0 \\ 0 & 1+\zeta & 0 \\ 0 & 0 & 1+\zeta \end{bmatrix} \tag{4-109}$$

The results of eigenvalue examination are given in Table 10.

Table 9 Eigenvalue of Stiffness Matrix for 4-node Degenerated Hybrid Element with multiplying matrix $[T_1]$ and E=1100 GPa, ν=0.3

No.	λ_i (*10^3)	No.	λ_i (*10^3)	No.	λ_i (*10^3)
1	0.2350	7	0.6214	13	1.4100
2	0.2637	8	0.6214	14	1.4100
3	0.2742	9	1.0060	15	1.5670
4	0.3626	10	1.0480	16	1.6920
5	0.3626	11	1.4100	17	1.6920
6	0.5641	12	1.4100	18	4.9510

Table 10 Eigenvalue of Stiffness Matrix for 4-node Degenerated Hybrid Element with multiplying matrix $[T_2]$ and E=1100 GPa, ν=0.3

No.	λ_i (*10^3)	No.	λ_i (*10^3)	No.	λ_i (*10^3)
1	0.1838	7	0.5668	13	1.5220
2	0.2115	8	0.5668	14	1.5220
3	0.3262	9	0.9530	15	1.6160
4	0.3441	10	0.9826	16	1.6920
5	0.3441	11	1.2480	17	1.6920
6	0.5641	12	1.2480	18	4.7500

The examination of the element shows that there are not any kinematic deformation modes when the assumed partial stress matrices (4-107)-(4-109) are used.

c) 8-node Partial Hybrid Degenerated Plate Element

An 8-node degenerated plate element [4.25-4.26] is also presented here using partial hybrid element formulation.

Geometry of Element

Firstly, the global co-ordinates (x,y,z) of any point within the element are expressed in the form specified by the 'vector' connecting the upper and lower points (see figure 19) and the mid-surface co-ordinates as

$$\begin{Bmatrix} x \\ y \\ z \end{Bmatrix} = \sum_{i=1}^{8} N_i(\xi,\eta) \begin{Bmatrix} x_i \\ y_i \\ z_i \end{Bmatrix} + \sum_{i=1}^{8} N_i(\xi,\eta)\, h_i \,\frac{\zeta}{2} V_{3i} \tag{4-110}$$

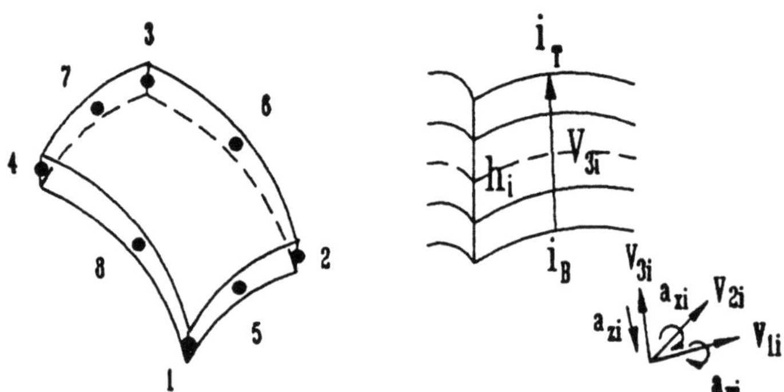

(a) (b)

Figure 19 An 8-node degenerated plate element

where (x_i, y_i, z_i) are the global co-ordinates of the i-th node. The shape functions $N_i(\xi,\eta)$ are

$$N_i=\frac{1}{4}(1+\xi_0)(1+\eta_0)(\xi_0+\eta_0-1)\xi_i^2\eta_i^2 \\ +\frac{1}{2}(1-\xi^2)(1+\eta_0)(1-\xi_i^2)\eta_i^2 \\ +\frac{1}{2}(1-\eta^2)(1+\xi_0)(1-\eta_i^2)\xi_i^2 \tag{4-111}$$

in which,

$$\xi_0=\xi_i\xi \quad \eta_0=\eta_i\eta \quad (i=1,2,\ldots,8) \tag{4-112}$$

The vector connecting the upper and lower points is

$$\mathbf{V}_{3i}=\begin{Bmatrix} l_{3i} \\ m_{3i} \\ n_{3i} \end{Bmatrix}=\frac{1}{h_i}\left(\begin{Bmatrix} x_i \\ y_i \\ z_i \end{Bmatrix}_T-\begin{Bmatrix} x_i \\ y_i \\ z_i \end{Bmatrix}_B\right) \tag{4-113}$$

and the parameter h_i is

$$h_i=\sqrt{(x_{iT}-x_{iB})^2+(y_{iT}-y_{iB})^2+(z_{iT}-z_{iB})^2} \tag{4-114}$$

Displacement Field

Similar to the 4-node degenerated element, it is assumed that a line that is straight and normal to the middle surface of the element before deformation is still straight, but not necessarily 'normal' to the middle surface after deformation. Therefore, a displacement field is assumed as

$$\begin{Bmatrix} u \\ v \\ w \end{Bmatrix}=\sum_{i=1}^{8} N_i\left[\begin{Bmatrix} u_i \\ v_i \\ w_i \end{Bmatrix}+\frac{\zeta}{2}h_i\,[\mathbf{V}_{1i}\ \mathbf{V}_{2i}\ \mathbf{V}_{3i}]\begin{Bmatrix} a_{xi} \\ a_{yi} \\ a_{zi} \end{Bmatrix}\right] \tag{4-115}$$

in which, $\mathbf{V}_{1i}$, $\mathbf{V}_{2i}$ and $\mathbf{V}_{3i}$ are the unit vectors of the local co-ordinate (ξ, η, ζ) at node i. They can be calculated as follows:

$$\boldsymbol{V}_{1i}=\begin{Bmatrix} l_{1i} \\ m_{1i} \\ n_{1i} \end{Bmatrix}=\frac{\boldsymbol{i}\times\boldsymbol{V}_{3i}}{|\boldsymbol{i}\times\boldsymbol{V}_{3i}|} \qquad \boldsymbol{V}_{2i}=\begin{Bmatrix} l_{2i} \\ m_{2i} \\ n_{2i} \end{Bmatrix}=\boldsymbol{V}_{3i}\times\boldsymbol{V}_{1i} \tag{4-116}$$

The displacement field can be rewritten in the same form as that for the 4-node degenerated plate element,

$$\begin{Bmatrix} u \\ v \\ w \end{Bmatrix}=\sum_{i=1}^{8} N_i\left[\begin{Bmatrix} u_i \\ v_i \\ w_i \end{Bmatrix}+\zeta\,[b_i]\begin{Bmatrix} a_{xi} \\ a_{yi} \\ a_{zi} \end{Bmatrix}\right]=\sum_{i=1}^{8}\,[N]_i\boldsymbol{\delta}_i \tag{4-117}$$

where

$$[b_i]=\begin{bmatrix} b_{11i} & b_{12i} & b_{13i} \\ b_{21i} & b_{22i} & b_{23i} \\ b_{31i} & b_{32i} & b_{33i} \end{bmatrix}=\frac{h_i}{2}\,[\boldsymbol{V}_{1i}\ \ -\boldsymbol{V}_{2i}\ \ \boldsymbol{V}_{3i}] \tag{4-118}$$

$$\boldsymbol{\delta}_i=[u_i\ v_i\ w_i\ a_{xi}\ a_{yi}\ a_{zi}]^T \tag{4-119}$$

$$[N]_i=\begin{bmatrix} N_i & 0 & 0 & N_i\zeta b_{11i} & N_i\zeta b_{12i} & N_i\zeta b_{13i} \\ 0 & N_i & 0 & N_i\zeta b_{21i} & N_i\zeta b_{22i} & N_i\zeta b_{23i} \\ 0 & 0 & N_i & N_i\zeta b_{31i} & N_i\zeta b_{32i} & N_i\zeta b_{33i} \end{bmatrix} \tag{4-120}$$

Partial Strain Field and Partial Derivatives of the Displacement Field

The partial globally continuous strains can be derived from the displacement field by means of partial strain-displacement relation as follows,

$$\boldsymbol{\varepsilon}_g=\begin{Bmatrix}\varepsilon_x\\ \varepsilon_y\\ \varepsilon_{xy}\end{Bmatrix}=\boldsymbol{D}_g\boldsymbol{u}=\begin{Bmatrix}\dfrac{\partial u}{\partial x}\\ \dfrac{\partial v}{\partial y}\\ \dfrac{\partial u}{\partial y}+\dfrac{\partial v}{\partial x}\end{Bmatrix}=[B_g]\boldsymbol{\delta} \tag{4-121}$$

in which,

$$[B_g]=[B_{g1}\ B_{g2}\ \ldots\ B_{g8}] \tag{4-122}$$

and

$$\boldsymbol{\delta}=[\boldsymbol{\delta}_1\ \boldsymbol{\delta}_2\ \ldots\ \boldsymbol{\delta}_8]^T \tag{4-123}$$

The expression of the partial geometric matrix at the i-th node $[B_{gi}]$ is the same as that of the 4-node degenerated plate element (4-85)-(4-86).

The partial derivatives of displacement field can be also derived from the displacement field as follows,

$$\boldsymbol{D}_L\boldsymbol{u}=\begin{Bmatrix}\dfrac{\partial w}{\partial z}\\ \dfrac{\partial v}{\partial z}+\dfrac{\partial w}{\partial y}\\ \dfrac{\partial w}{\partial x}+\dfrac{\partial u}{\partial z}\end{Bmatrix}=[B_L]\boldsymbol{\delta} \tag{4-124}$$

in which,

$$[B_L]=[B_{L1}\ B_{L2}\ \ldots\ B_{L8}] \tag{4-125}$$

The expression of the partial geometric matrix at the i-th node $[B_{Li}]$ is the same as that of the 4-node degenerated plate element (4-89).

In order to calculate $N_{i,x}$, $N_{i,y}$, $N_{i,z}$ and $\zeta_{,x}$, $\zeta_{,y}$, $\zeta_{,z}$, the following vectors are introduced:

$$\boldsymbol{S}=\begin{Bmatrix} x_{,\xi} \\ y_{,\xi} \\ z_{,\xi} \end{Bmatrix}=\sum_{i=1}^{8} N_{i,\xi}\left(\begin{Bmatrix} x_i \\ y_i \\ z_i \end{Bmatrix}+\frac{h_i}{2}\zeta \boldsymbol{V_{3i}}\right) \tag{4-126}$$

$$\boldsymbol{T}=\begin{Bmatrix} x_{,\eta} \\ y_{,\eta} \\ z_{,\eta} \end{Bmatrix}=\sum_{i=1}^{8} N_{i,\eta}\left(\begin{Bmatrix} x_i \\ y_i \\ z_i \end{Bmatrix}+\frac{h_i}{2}\zeta \boldsymbol{V_{3i}}\right) \tag{4-127}$$

$$\boldsymbol{V}=\begin{Bmatrix} x_{,\zeta} \\ y_{,\zeta} \\ z_{,\zeta} \end{Bmatrix}=\sum_{i=1}^{8} N_i\frac{h_i}{2}\boldsymbol{V_{3i}} \tag{4-128}$$

Thus, the Jacobian matrix, the derivatives of the shape functions, and the geometric matrix [B'] in the local co-ordinate system can be obtained using equations (4-93)-(4-103).

Assumed Partial Stress Field

In the element, the partial stress field is assumed independently as continuous functions along the thickness of a composite structure.

$$\boldsymbol{\sigma_g}=\begin{Bmatrix} \sigma_z \\ \tau_{yz} \\ \tau_{zx} \end{Bmatrix}=[P_g]\,\boldsymbol{\beta}=[T]\,[P]\,\boldsymbol{\beta} \tag{4-129}$$

where the multiplying matrix [T] is determined by the traction conditions on the top and bottom surfaces of the structure. The matrix [P] is derived directly from the assumed displacement field using the iso-function method. The iso-function partial stress matrix of the element is

$$
[P] = \left| \begin{array}{cccccccccccccccccc}
1 & 0 & 0 & \xi & 0 & 0 & \eta & 0 & 0 & \zeta & 0 & 0 & \xi\eta & 0 & 0 & \xi\zeta & 0 & 0 \\
0 & 1 & 0 & 0 & \xi & 0 & 0 & \eta & 0 & 0 & \zeta & 0 & 0 & \xi\eta & 0 & 0 & \xi\zeta & 0 \\
0 & 0 & 1 & 0 & 0 & \xi & 0 & 0 & \eta & 0 & 0 & \zeta & 0 & 0 & \xi\eta & 0 & 0 & \xi\zeta
\end{array} \right.
$$

$$
\begin{array}{ccccccccccccc}
\eta\zeta & 0 & 0 & \xi\eta\zeta & 0 & 0 & \xi^2 & 0 & 0 & \eta^2 & 0 & 0 & \xi^2\eta \\
0 & \eta\zeta & 0 & 0 & \xi\eta\zeta & 0 & 0 & \xi^2 & 0 & 0 & \eta^2 & 0 & 0 \\
0 & 0 & \eta\zeta & 0 & 0 & \xi\eta\zeta & 0 & 0 & \xi^2 & 0 & 0 & \eta^2 & 0
\end{array} \qquad (4\text{-}130)
$$

$$
\left. \begin{array}{ccccccccc}
0 & 0 & \xi^2\zeta & 0 & \xi\eta^2 & 0 & 0 & \eta^2\zeta & 0 \\
\xi^2\eta & 0 & 0 & \xi^2\zeta & 0 & \xi\eta^2 & 0 & 0 & 0 \\
0 & \xi^2\eta & 0 & 0 & 0 & 0 & \xi\eta^2 & 0 & \eta^2\zeta
\end{array} \right|
$$

There are 40 stress modes in the iso-function partial stress matrix.

Examination of Partial Hybrid Element

In the element, there are 8 nodes and each node has six components of displacements. Therefore, the element has (n=) 48 degrees of freedom. For a 3-D elastic body, the degrees of the rigid displacement are equal to (r=) 6. Thus, the element has (n-r=) 42 natural deformation modes. The eigenvalue examination indicates that the partial stiffness matrix $[K_d]$ has 26 non-zero eigenvalues. So the rank of the partial stiffness matrix $[K_d]$ is (n_d=) 26 and the matrix $[K_d]$ represents 26 natural deformation modes. Thus, another partial stiffness matrix $[K_h]$ must represent 16 natural deformation modes of the element and the assumed partial stress matrix must contain at least 16 stress modes.

Based on the iso-function partial stress matrix (4-130), when the multiplying matrix [T] is a unit matrix [I], the classification method gives an optimal stress matrix as follows,

$$
[P] = \begin{bmatrix}
1 & 0 & 0 & \xi & 0 & 0 & \eta & 0 & 0 & \xi\eta & 0 & 0 & 0 & 0 & 0 & 0 \\
0 & 1 & 0 & 0 & \xi & 0 & 0 & \eta & 0 & 0 & \xi\eta & 0 & \xi\zeta & 0 & \xi\eta\zeta & 0 \\
0 & 0 & 1 & 0 & 0 & \xi & 0 & 0 & \eta & 0 & 0 & \xi\eta & 0 & \eta\zeta & 0 & \xi\eta\zeta
\end{bmatrix} \qquad (4\text{-}131)
$$

The degenerated plate elements with three types of materials are examined. The first is for isotropic material (see Table 11); the second is for anisotropic material (see Table 12); The third is for the composite structure with fibre orientation [90, 0, 90] (see Table 13). In the tables,

$$\lambda_i = \frac{\lambda_{hi}}{\lambda_{ui}} \tag{4-132}$$

where λ_{hi} is the eigenvalue of the hybrid element; λ_{ui} is the eigenvalue of the conventional displacement element. There are not any spurious zero eigenvalues. Therefore, the elements do not have any kinematic deformation modes.

Table 11 Eigenvalue Analysis of Stiffness Matrix for the Degenerated Element with 16 stress modes and isotropic materials: E=1100 GPa, ν=0.3

No.	λ_i	No.	λ_i	No.	λ_i
1	0.7355	15	0.9253	29	0.9276
2	0.9392	16	0.9253	30	0.8042
3	0.7040	17	0.9241	31	1.0000
4	0.6753	18	0.7388	32	0.9966
5	0.8229	19	0.9190	33	0.9994
6	0.5870	20	0.9762	34	0.9994
7	0.6636	21	0.9790	35	0.9817
8	0.6092	22	0.8131	36	0.9841
9	0.7219	23	0.7528	37	0.8384
10	0.8699	24	0.7375	38	0.8432
11	0.8197	25	0.8694	39	0.9995
12	0.8974	26	0.8120	40	0.9114
13	0.8793	27	0.8195	41	1.0000
14	0.8793	28	0.8195	42	1.0000

Table 12 Eigenvalue Analysis of Stiffness Matrix for the Degenerated Element with 16 stress modes and anisotropic materials: E_L=174.6 GPa, E_T=7.0 GPa, G_{LT}=3.5 GPa, G_{TT}=1.4 GPa, $\nu_{12}=\nu_{13}=\nu_{23}$=0.25

No.	λ_i	No.	λ_i	No.	λ_i
1	0.9474	15	0.9094	29	0.9766
2	0.4623	16	0.8548	30	1.0000
3	0.5917	17	0.7773	31	0.9993
4	0.7961	18	0.7797	32	0.8925
5	0.8069	19	0.9696	33	0.9897
6	0.7233	20	0.8894	34	0.9968
7	0.7416	21	0.9453	35	0.9998
8	0.7526	22	0.9086	36	1.0000
9	0.8614	23	0.9842	37	0.9940
10	0.8387	24	0.8846	38	0.9918
11	0.8810	25	0.9896	39	0.9995
12	0.8917	26	0.9820	40	1.0000
13	0.9025	27	0.9996	41	0.9916
14	0.9138	28	0.9496	42	1.0000

Table 13 Eigenvalue Analysis of Stiffness Matrix for the Degenerated Element with fibre orientation [90, 0,90], 16 stress modes and materials: E_L=174.6 GPa, E_T=7.0 GPa, G_{LT}=3.5 GPa, G_{TT}=1.4 GPa, $\nu_{12}=\nu_{13}=\nu_{23}$=0.25

No.	λ_i	No.	λ_i	No.	λ_i
1	0.6746	15	0.8201	29	0.9094
2	0.5697	16	0.7790	30	0.9506
3	0.5951	17	0.7862	31	0.9982
4	0.7661	18	0.9046	32	1.0000
5	0.8326	19	0.9080	33	0.9999
6	0.7248	20	0.8470	34	0.9899
7	0.6863	21	0.8440	35	0.9915
8	0.6293	22	0.7805	36	0.9986
9	0.7808	23	0.9788	37	1.0000
10	0.8043	24	1.0000	38	0.9993
11	0.8668	25	0.9470	39	1.0000
12	0.7016	26	0.9717	40	0.9896
13	0.8314	27	0.8866	41	0.9998
14	0.8170	28	0.8840	42	0.9999

4.2.4 Partial Hybrid Transition Element

In practical applications, a number of the so-call 'global/local' solution method have been proposed [4.27] in order to determine the stress state in composite structures efficiently and accurately using 3-D finite element model. In the global-local analysis, the treatment of interfaces between global and local regions is one of the key elements. One of the commonly-used approaches for maintaining displacement compatibility and traction reciprocity at the interfaces is a special transition element. The major advantage of the transition element is to eliminate the constraint equations at these transition regions[4.28-4.30]. Two partial hybrid transition elements are presented here [4.31-4.36]. They will be used to connect the 3-D partial hybrid solid elements in local region with the partial hybrid degenerated plate elements in global region for the global/local analysis of composite structures as presented in chapter 5.

a) 6-node Partial Hybrid Transition Element

This transition element is used to connect partial hybrid degenerated plate elements and 3-D partial hybrid solid elements. It has two line of nodes[4.37] where it meets the degenerated plate element and four point nodes on the remaining boundaries where it meets the 3-D solid element (see Figure 20). The line of nodes can accommodate any function along the thickness, allowing it to admit the any-order polynomials over the entire thickness from degenerated plate element, while the point nodes have the same polynomial shape functions as those used for the 3-D solid element. Once the shape functions of the transition element are established, the definitions of geometry and displacements for the element follow a similar path to those of the hybrid elements.

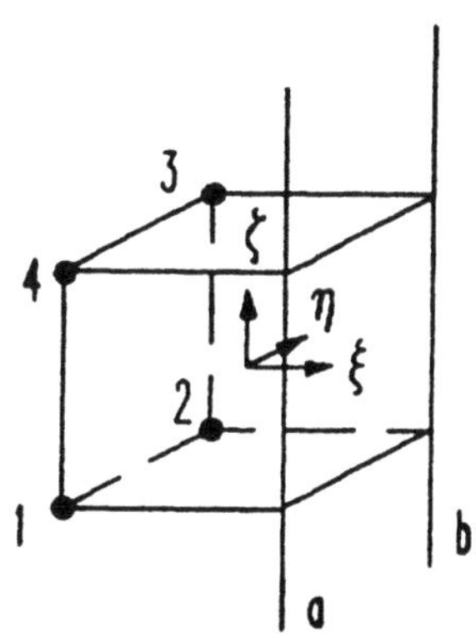

Figure 20 6-node partial hybrid transition element

The Shape Functions of Different Elements

In the formulation of a typical 3-D element as shown in figure 21(a), shape functions dictate the form of any field, e.g. displacement fields. It is clear that if two adjacent elements have identical shape functions and nodal locations on the interface, the continuity of any field between the elements is achieved and the elements are compatible. Otherwise, the compatibility of elements will not be satisfied.

Suppose that a transition element in Figure 21(b) is used to connect a solid element (Fig. 21a) to a plate element (Fig. 21c) in the transition region of a global/local finite element model. On the left side, it meets with a solid element; on the right side, it meets a plate element. Thus, the transition element must have the same shape functions and nodal locations on its left side as that of solid element, and on its right side as that of plate/shell element.

Take a solid element as an "original" element for developing the transition element. Obviously, the shape functions on the left side do not need to be modified. But, on the right side, new shape functions for satisfying continuity are required. For a general case, this amounts to developing a set of shape functions which can accommodate any arbitrary curve specified by the adjacent right-side element along the ζ axis (thickness) on the interface between elements. Before attempting to generate such shape functions, it is instructive to examine shape functions for a solid element.

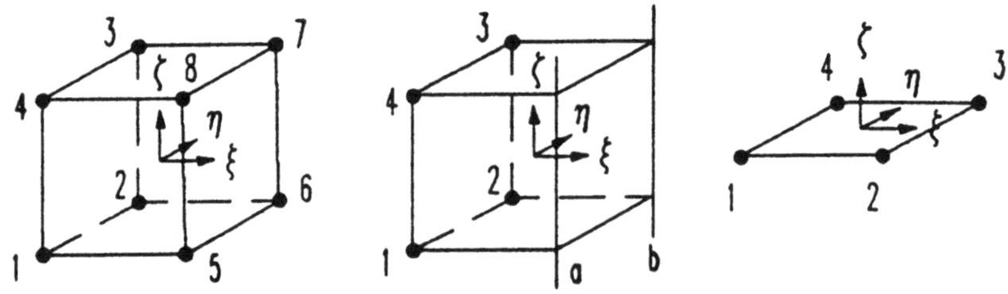

(a) solid element (b) transition element (c) plate element

Figure 21. Three types of element

In figure 21(a), a typical linear solid element is shown with the local curvilinear co-ordinates ξ, η and ζ. Its shape functions can be found elsewhere[4.38-4.39]. For developing a transition element, of particular interest are the shape functions for the nodes on the right side of this element, node 5-8:

$$N_i = \frac{1}{8}(1+\xi_i\xi)(1+\eta_i\eta)(1+\zeta_i\zeta) \tag{4-133}$$

The transition element in Fig. 21(b) is similar to the solid element in Fig 21(a) except for the nodes on its right face (5,6,7,8). Special treatment has to be done to these nodes so that their displacements can be compatible to those of the plate element in figure 21(c). Consider a function,

$$\Omega(\xi,\eta,\zeta) = \Omega_a(\xi,\eta,\zeta) + \Omega_b(\xi,\eta,\zeta) \tag{4-134}$$

in which,

$$\begin{aligned} \Omega_a(\xi,\eta,\zeta) &= \alpha_5 N_5 + \alpha_8 N_8 \\ \Omega_b(\xi,\eta,\zeta) &= \alpha_6 N_6 + \alpha_7 N_7 \end{aligned} \tag{4-135}$$

where α_i is the value of Ω at the node i of the solid element. Note that the function Ω can have the meaning of displacement function for the right face of the transition element in Fig. 21(b). Functions Ω_a and Ω_b can be of any degree (linear, quadratic etc.) between the two nodes 5-8 or 6-7 respectively. For a regular brick element, α_5 and α_8 would represent the displacements at the node 5 and 8 respectively. Normally, the displacements at node 5 and 8 are independent from each other. However, if the displacements at nodes 5 and 8 are constrained such that each of them is equal to a specific value of a function $\beta(\zeta')$, then one can write:

$$\begin{aligned} \alpha_5 &= \beta(\zeta'_5) \\ \alpha_8 &= \beta(\zeta'_8) \end{aligned} \tag{4 - 136}$$

where ζ'_5 and ζ'_8 are the global coordinates in the thickness direction of nodes 5 and 8 respectively.

Now consider the shape functions of the degenerated plate element. A middle surface of a degenerated plate element is shown with the local curvilinear co-ordinates ξ and η (in Figure 21(c)). The shape functions for the nodes on the left boundary of this element, nodes 1 and 4 are:

$$\overline{N_1}=(1-\xi)(1-\eta)/4$$
$$\overline{N_4}=(1-\xi)(1+\eta)/4 \tag{4-137}$$

Consider another function meantime,

$$\Pi(\xi,\eta,\zeta')=\Pi_1(\xi,\eta,\zeta')+\Pi_4(\xi,\eta,\zeta') \tag{4-138}$$

in which

$$\Pi_1(\xi,\eta,\zeta')=A_1\beta_1(\zeta')\overline{N_1}$$
$$\Pi_4(\xi,\eta,\zeta')=A_4\beta_4(\zeta')\overline{N_4} \tag{4-139}$$

and Π represents the displacements at the left face of the plate element. Note that the function Π can be considered to be the displacements of points lying on a plane normal to the middle surface of the plate element at edge 1-4. Π_1 can be considered to be the displacements at all points on the normal to the initial middle surface of the undeformed plate element at node 1 and Π_4 can be considered likewise to be the displacements at all points on the normal to the initial middle surface of the undeformed plate element at node 4. Let us consider the composition of Π_1 in detail. The composition of Π_4 follows.

In Π_1 , $\overline{N}_1$ represents the shape function in the plane ξ-η. A_1 represents the nodal displacement at the node 1 on the middle surface of the plate element. $\beta_1(\zeta')$ represents the variation of the displacement in the undeformed state of any point initially lying on the line normal to the middle surface of the plate element. If only one plate element is used for the whole laminate thickness, $\beta(\zeta')$ is a linear function of ζ'.

Matching the Two Shape Functions

From figure 21(b), Ω represents the displacement of the transition element at the interface. From figure 21(c), Π represents the displacement of the plate element at the interface. In order to satisfy the compatibility of displacement fields at the interface between the transition element and plate element, Ω and Π must be the same.

Function Ω consists of two functions Ω_a and Ω_b, and function Π consists of another two functions Π_1 and Π_4. At the interface, Ω_a and Ω_b will need to match Π_1 and Π_4, respectively. If Ω_i and Π_i can be matched exactly, compatibility of displacement fields at the interface will be satisfied. At the interface, ζ' is a function of ζ. In the interval $\zeta \in [-1,1]$ which corresponds to $\zeta' \in [\zeta'_l, \zeta'_{l+1}]$, one has

$$\zeta' = \frac{1-\zeta}{2}\zeta'_l + \frac{1+\zeta}{2}\zeta'_{l+1} \tag{4-140}$$

Note that ζ' is the global thickness coordinate for the plate element while ζ is the smaller thickness coordinate for the solid or transition element. Using equation (4-133), the function Ω_a in equation (4-135) can be rewritten as follows:

$$\Omega_a = \frac{1}{8}(1+\xi)(1-\eta)[(1-\zeta)\alpha_5 + (1+\zeta)\alpha_8] \tag{4-141}$$

In this expression, the function Ω_a is split into two parts $N_a(\xi,\eta)$ and $\alpha_a(\zeta)$ that are

$$N_a(\xi,\eta) = \frac{1}{4}(1+\xi)(1-\eta)$$
$$\alpha_a(\zeta) = \frac{1}{2}[(1-\zeta)\alpha_5 + (1+\zeta)\alpha_8] \tag{4-142}$$

Thus,

$$\Omega_a = \alpha_a N_a \tag{4-143}$$

In order to accommodate any arbitrary curve $\beta_1(\zeta')$ specified by the adjoining plate element, a line of nodes connecting nodes 5-8 and a moving node which moves along this line are defined. At every point ζ' occupied by the moving node, the nodal value α_a is made to be equal to the value of the specified curve at the point, $A_1\beta_1(\zeta')$. Thus,

$$\alpha_a = A_1\beta_1(\zeta')$$
$$\Omega_a = A_1\beta_1(\zeta')N_a(\xi,\eta) \tag{4-144}$$

Because the contribution of line 5-8 to the displacement field of the transition element is represented by the function Ω_a, this line is called as line of nodes "a" in order to use the standard word "node" in finite element method. Now, comparing functions Ω_a and Π_1 (4-139) and (4-144) at the interface, one can see that the functions Ω_a and Π_1 are the same (note that $N_a=\overline{N}_1$ at the interface). The Ω_a and Π_1 are matched exactly. In the same way as the functions Ω_a and Π_1, The Ω_b and Π_4 can also be matched. The new shape functions and nodal values are defined by

$$\alpha_b=A_4\beta_4(\zeta') \qquad \text{(4-145)}$$

$$N_b=(1+\xi)(1+\eta)/4$$

Thus

$$\Omega_b=\alpha_b N_b=A_4\beta_4(\zeta')N_b(\xi,\eta) \qquad \text{(4-146)}$$

The functions Ω_a and Ω_b of two lines of nodes "a" and "b" determine the displacements of the transition element at interface. Similarly, the functions Π_1 and Π_4 of two nodes 1 and 4 determine the displacements of the plate element at the interface. Because the functions Ω_a and Ω_b are the same as the functions Π_1 and Π_4 at the interface respectively, the function Ω is subsequently same as the function Π. Therefore, the displacements are compatible at the interface between the transition element and plate element.

These two new shape functions N_a and N_b along with the other four shape functions as given in Reference [4.38-4.39] form a complete set of shape functions for the transition element.

Geometry of the Element

A transition element is shown in Figure 22. The global co-ordinate (x, y, z) of any point in the element may be related to the non-dimensional co-ordinates by

$$\begin{Bmatrix} x \\ y \\ z \end{Bmatrix}=\sum_{1}^{4} N_i \begin{Bmatrix} x_i \\ y_i \\ z_i \end{Bmatrix}+\sum_{5}^{8} N_i \begin{Bmatrix} x_i \\ y_i \\ z_i \end{Bmatrix} \qquad \text{(4-147)}$$

in which, x_i, y_i and z_i are the co-ordinates of node i. Because the point nodes 5-8 are replaced by two lines of node a and b as follows,

$$\sum_{5}^{8} N_i \begin{Bmatrix} x_i \\ y_i \\ z_i \end{Bmatrix} = \sum_{a}^{b} N_i \left(\begin{Bmatrix} x_i^0 \\ y_i^0 \\ z_i^0 \end{Bmatrix} + \frac{\zeta'}{2} \begin{Bmatrix} \Delta x_i \\ \Delta y_i \\ \Delta z_i \end{Bmatrix} \right) \tag{4-148}$$

where, x_i^0, y_i^0 and z_i^0 (i=a,b) are the co-ordinates of the lines "a" and "b" at the middle surface of the composite structure, one has

$$\begin{Bmatrix} x \\ y \\ z \end{Bmatrix} = \sum_{1}^{4} N_i \begin{Bmatrix} x_i \\ y_i \\ z_i \end{Bmatrix} + \sum_{a}^{b} N_i \left(\begin{Bmatrix} x_i^0 \\ y_i^0 \\ z_i^0 \end{Bmatrix} + \frac{\zeta'}{2} \begin{Bmatrix} \Delta x_i \\ \Delta y_i \\ \Delta z_i \end{Bmatrix} \right) \tag{4-149}$$

Moreover, it can be rewritten in the form,

$$\begin{Bmatrix} x \\ y \\ z \end{Bmatrix} = \sum_{1}^{4} N_i \begin{Bmatrix} x_i \\ y_i \\ z_i \end{Bmatrix} + \sum_{a}^{b} N_i \left(\begin{Bmatrix} x_i^0 \\ y_i^0 \\ z_i^0 \end{Bmatrix} + \frac{\zeta'}{2} h_i \mathbf{V}_{3i} \right) \tag{4-150}$$

where

$$\mathbf{V}_{3i} = \begin{Bmatrix} l_{3i} \\ m_{3i} \\ n_{3i} \end{Bmatrix} = \frac{1}{h_i} \left(\begin{Bmatrix} x_i \\ y_i \\ z_i \end{Bmatrix}_T - \begin{Bmatrix} x_i \\ y_i \\ z_i \end{Bmatrix}_B \right) \tag{4-113}$$

and

$$h_i = \sqrt{(x_{iT} - x_{iB})^2 + (y_{iT} - y_{iB})^2 + (z_{iT} - z_{iB})^2} \tag{4-114}$$

The shape function N_i can be expressed as follows:

$$N_i = \frac{1}{8} (1+\xi_0)(1+\eta_0)(1+\zeta_0) \tag{4-151}$$

in which

$$\xi_0=\xi_i\xi \quad \eta_0=\eta_i\eta \quad \zeta_0=\zeta_i\zeta \qquad i=1-4 \tag{4-152}$$

and

$$N_a=(1+\xi)(1-\eta)/4$$
$$N_b=(1+\xi)(1+\eta)/4 \tag{4-153}$$

Note that coordinate ζ' is the global thickness coordinate for the plate element and coordinate ζ is the smaller thickness coordinate for the transition element. The relationship between coordinate ζ' and coordinate ζ is expressed in the equation (4-140). The values ζ'_1 and ζ'_{1+1} represent the values of co-ordinate ζ' at the lower and upper surfaces of a layer while ζ=-1 and ζ=+1, respectively.

Displacement Field

In the element, the displacements (see figure 22) are expressed as follows:

$$\begin{Bmatrix} U \\ V \\ W \end{Bmatrix}=\sum_{1}^{4} N_i \begin{Bmatrix} U_i \\ V_i \\ W_i \end{Bmatrix}+\sum_{a}^{b} N_i \left[\begin{Bmatrix} U_i^0 \\ V_i^0 \\ W_i^0 \end{Bmatrix}+\zeta' [b_i] \begin{Bmatrix} \Psi_{xi} \\ \Psi_{yi} \\ \Psi_{zi} \end{Bmatrix} \right] \tag{4-154}$$

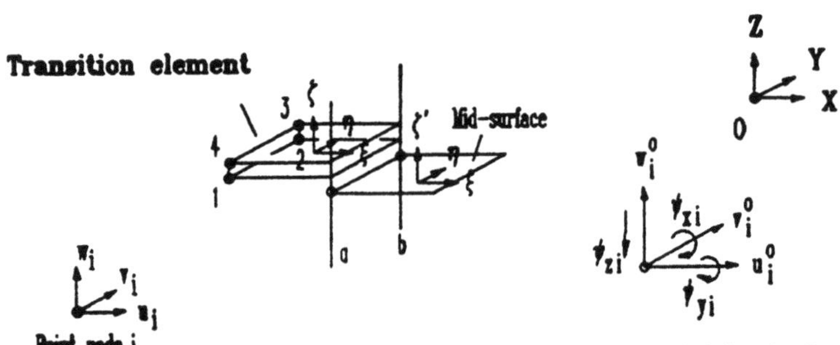

Figure 22 Nodal displacements in a transition sub-element

where

$$[b_i] = \begin{bmatrix} b_{11i} & b_{12i} & b_{13i} \\ b_{21i} & b_{22i} & b_{23i} \\ b_{31i} & b_{32i} & b_{33i} \end{bmatrix} = \frac{h_i}{2}\,[\mathbf{V}_{1i}\ \ -\mathbf{V}_{2i}\ \ \mathbf{V}_{3i}] \tag{4-118}$$

and

$$\mathbf{V}_{1i} = \begin{Bmatrix} l_{1i} \\ m_{1i} \\ n_{1i} \end{Bmatrix} = \frac{\mathbf{i}\times\mathbf{V}_{3i}}{|\mathbf{i}\times\mathbf{V}_{3i}|} \qquad \mathbf{V}_{2i} = \begin{Bmatrix} l_{2i} \\ m_{2i} \\ n_{2i} \end{Bmatrix} = \mathbf{V}_{3i}\times\mathbf{V}_{1i} \tag{4-116}$$

in which, the displacement components u_i , v_i and w_i are the nodal displacements at point nodes 1,2,3 and 4. The displacement components u_i^0, v_i^0 and w_i^0 are the displacements of the line "a" and "b" at the middle surface of the composite structure, ψ_{xi} and ψ_{yi} are two rotations of the nodal vector $\mathbf{V}_{3i}$ about orthogonal directions normal to it, and ψ_{zi} is a transverse normal deformation in the thickness direction.

Partial Strain Field and Partial Derivatives of the Displacement Field

The partial strain field is

$$\{\boldsymbol{\varepsilon}_g\} = \begin{Bmatrix} \dfrac{\partial u}{\partial x} \\ \dfrac{\partial v}{\partial y} \\ \dfrac{\partial u}{\partial y} + \dfrac{\partial v}{\partial x} \end{Bmatrix} = [B_g]\boldsymbol{\delta} = [B_{g1}B_{g2}\cdot\cdot\cdot\cdot\cdot B_{g6}] \begin{Bmatrix} \boldsymbol{\delta}_1 \\ \boldsymbol{\delta}_2 \\ \cdot \\ \cdot \\ \boldsymbol{\delta}_6 \end{Bmatrix} \tag{4-155}$$

where $\delta_5 = \delta_a$ and $\delta_6 = \delta_b$. The partial derivatives of the displacement are

$$\mathbf{D}_L\mathbf{u} = \begin{Bmatrix} \dfrac{\partial w}{\partial z} \\ \dfrac{\partial v}{\partial z} + \dfrac{\partial w}{\partial y} \\ \dfrac{\partial w}{\partial x} + \dfrac{\partial u}{\partial z} \end{Bmatrix} = [B_L]\boldsymbol{\delta} = \{B_{L1}B_{L2}\cdot\cdot\cdot\cdot\cdot B_{L6}] \begin{Bmatrix} \boldsymbol{\delta}_1 \\ \boldsymbol{\delta}_2 \\ \cdot \\ \cdot \\ \boldsymbol{\delta}_6 \end{Bmatrix} \tag{4-156}$$

For nodes i = 1 - 4,

$$[B_{gi}]=\begin{bmatrix} N_{i,x} & 0 & 0 \\ 0 & N_{i,y} & 0 \\ N_{i,y} & N_{i,x} & 0 \end{bmatrix} \tag{4-157}$$

and

$$[B_{Li}]=\begin{bmatrix} 0 & 0 & N_{i,z} \\ 0 & N_{i,z} & N_{i,y} \\ N_{i,z} & 0 & N_{i,x} \end{bmatrix} \tag{4-158}$$

and

$$\boldsymbol{\delta}_i=\begin{Bmatrix} u_i \\ v_i \\ w_i \end{Bmatrix} \tag{4-159}$$

For nodes i = a and b,

$$[B_{gi}] = \begin{bmatrix} N_{i,x} & 0 & 0 & b_{11i}a_{ix} & b_{12i}a_{ix} & b_{13i}a_{ix} \\ 0 & N_{i,y} & 0 & b_{21i}a_{iy} & b_{22i}a_{iy} & b_{23i}a_{iy} \\ N_{i,y} & N_{i,x} & 0 & b_{11i}a_{iy}+b_{21i}a_{ix} & b_{12i}a_{iy}+b_{22i}a_{ix} & b_{13i}a_{iy}+b_{23i}a_{ix} \end{bmatrix} \tag{4-160}$$

and

$$[B_{Li}] = \begin{bmatrix} 0 & 0 & N_{i,z} & b_{31i}a_{iz} & b_{32i}a_{iz} & b_{33i}a_{iz} \\ 0 & N_{i,z} & N_{i,y} & b_{21i}a_{iz}+b_{31i}a_{iy} & b_{22i}a_{iz}+b_{32i}a_{iy} & b_{23i}a_{iz}+b_{33i}a_{iy} \\ N_{i,z} & 0 & N_{i,x} & b_{31i}a_{ix}+b_{11i}a_{iz} & b_{32i}a_{ix}+b_{12i}a_{iz} & b_{33i}a_{ix}+b_{13i}a_{iz} \end{bmatrix}$$

(4-161)

and

(4-162)

$$\boldsymbol{\delta}_i=[u_i^0 \;\; v_i^0 \;\; w_i^0 \;\; \psi_{xi} \;\; \psi_{yi} \;\; \psi_{zi}]^T$$

and

$$
\begin{aligned}
a_{ix} &= N_{i,x}\zeta' + N_i\zeta'_{,x} \\
a_{iy} &= N_{i,y}\zeta' + N_i\zeta'_{,y} \\
a_{iz} &= N_{i,z}\zeta' + N_i\zeta'_{,z}
\end{aligned}
\tag{4-163}
$$

In order to calculate $N_{i,x}$, $N_{i,y}$, $N_{i,z}$ and $\zeta'_{,x}$, $\zeta'_{,y}$, $\zeta'_{,z}$, the following vectors are introduced:

$$
\boldsymbol{r}=\begin{pmatrix} x \\ y \\ z \end{pmatrix} \qquad \boldsymbol{S}=\boldsymbol{r}_{,\xi}=\begin{pmatrix} x_{,\xi} \\ y_{,\xi} \\ z_{,\xi} \end{pmatrix}
\tag{4-164}
$$

$$
\boldsymbol{T}=\boldsymbol{r}_{,\eta}=\begin{pmatrix} x_{,\eta} \\ y_{,\eta} \\ z_{,\eta} \end{pmatrix} \qquad \boldsymbol{V}=\boldsymbol{r}_{,\zeta}=\begin{pmatrix} x_{,\zeta} \\ y_{,\zeta} \\ z_{,\zeta} \end{pmatrix}
$$

Then

$$
[J]=[\boldsymbol{S}\ \boldsymbol{T}\ \boldsymbol{V}]^T \quad \textit{and} \quad |J|=\boldsymbol{S}\cdot\boldsymbol{T}\times\boldsymbol{V}
\tag{4-165}
$$

and

$$
[J]^{-1}=[\boldsymbol{T}\times\boldsymbol{V}\ \ \boldsymbol{V}\times\boldsymbol{S}\ \ \boldsymbol{S}\times\boldsymbol{T}]/|J|
\tag{4-166}
$$

One can obtain

$$
\begin{Bmatrix} N_{i,x} \\ N_{i,y} \\ N_{i,z} \end{Bmatrix} = [\boldsymbol{T}\times\boldsymbol{V}\ \ \boldsymbol{V}\times\boldsymbol{S}\ \ \boldsymbol{S}\times\boldsymbol{T}]\ /|J| \begin{Bmatrix} N_{i,\xi} \\ N_{i,\eta} \\ N_{i,\zeta} \end{Bmatrix}
\tag{4-167}
$$

and

$$
\begin{Bmatrix} \zeta_{,x} \\ \zeta_{,y} \\ \zeta_{,z} \end{Bmatrix} = \frac{\boldsymbol{S}\times\boldsymbol{T}}{|J|}
\tag{4-168}
$$

Assumed Partial Stress Field

In the element, the partial stress field is also assumed independently.

$$\boldsymbol{\sigma}_g = \begin{Bmatrix} \sigma_z \\ \tau_{yz} \\ \tau_{zx} \end{Bmatrix} = [P_g]\,\boldsymbol{\beta} \tag{4-169}$$

where the stress parameters β_i are the internal parameters. In some cases, it is convenient to use surface stress parameters α. For example, an assumed stress field can be assumed in the form,

$$\boldsymbol{\sigma}_g = [P_g]\,\boldsymbol{\beta} = [P]\,\frac{1}{2}\{(1+\zeta)\,\boldsymbol{\alpha}_T + (1-\zeta)\,\boldsymbol{\alpha}_B\} \tag{4-170}$$

where α_T and α_B are the surface stress parameters corresponding to upper and lower surfaces of the element, respectively. In this expression, a stress mode σ_i in the matrix [P] is related to both surfaces (upper and lower surfaces α_T and α_B) and corresponds two stress modes $0.5*(1+\zeta)*\sigma_i$ and $0.5*(1-\zeta)*\sigma_i$ in the assumed stress matrix $[P_g]$. The stress matrix [P] can be derived directly from the assumed displacement field using the iso-function method. The iso-function partial stress matrix of the element is

$$[P_g] = \begin{bmatrix} 1 & 0 & 0 & \xi & 0 & 0 & \eta & 0 & 0 & \zeta & 0 & 0 & \xi\eta & 0 & 0 & \eta\zeta & 0 & \xi\zeta & 0 \\ 0 & 1 & 0 & 0 & \xi & 0 & 0 & \eta & 0 & 0 & \zeta & 0 & 0 & \xi\eta & 0 & 0 & 0 & 0 & \xi\zeta \\ 0 & 0 & 1 & 0 & 0 & \xi & 0 & 0 & \eta & 0 & 0 & \zeta & 0 & 0 & \xi\eta & 0 & \eta\zeta & 0 & 0 \end{bmatrix} \tag{4-171}$$

There are 19 stress modes in the stress matrix.

Examination of Partial Hybrid Transition Element

In this element, there are four point nodes and two lines of nodes. Each point node has three components of displacements and each line of nodes has six components. Therefore, the element has (n=) 24 degrees of freedom. Because the degrees of the rigid body motion are equal to (r=) 6, the element has (n-r=) 18 natural deformation modes. The eigenvalue examination indicates that the rank of the partial stiffness matrix $[K_d]$ for the element is (n_d=) 10. Therefore, the partial stiffness matrix $[K_d]$ represents 10 natural deformation modes of the element. Another partial stiffness matrix $[K_h]$ must represent 8 natural deformation modes of the element and the assumed stress matrix $[P_g]$ must contain at least 8 stress modes. Due to the fact that a stress mode in the matrix [P] corresponds two stress modes in the assumed stress matrix $[P_g]$, the stress matrix [P] must have at least 4 stress modes.

Based on the iso-function partial stress matrix (4-171), the classification method gives an optimal stress matrix as follows,

$$[P] = \begin{bmatrix} 1 & 0 & 0 & 0 & 0 \\ 0 & 1 & 0 & \xi & 0 \\ 0 & 0 & 1 & 0 & \eta \end{bmatrix} \tag{4-172}$$

The examination of the element shows that there are not any kinematic deformation modes. The results of eigenvalue examination are given in Table 14. In the table, λ_i are the eigenvalues of the element.

Table 14 Eigenvalues of Stiffness Matrix for Hybrid Transition Element with 10 stress modes and isotropic materials: E=1100 GPa, ν=0.3

No.	λ_i (*10^3)	No.	λ_i (*10^3)	No.	λ_i (*10^3)
1	0.1088	7	0.2923	13	1.282
2	0.1621	8	0.4328	14	1.373
3	0.1850	9	0.6554	15	1.398
4	0.2115	10	0.7126	16	1.398
5	0.2209	11	0.7264	17	1.418
6	0.2438	12	0.9706	18	4.171

b) 15-node Partial Hybrid Transition Element

In order to connect 8-node partial hybrid degenerated plate elements and 3-D, 20-node partial hybrid solid elements, a 15-node transition element is presented. It has three lines of nodes [4.37] where they meet the degenerated plate/shell element and four point nodes on the remaining boundaries (see Figure 23). The line of nodes can accommodate any function along the thickness, allowing it to admit the any-order polynomials over the entire thickness from degenerated plate element, while the point nodes have the same polynomial shape functions as those used for the 3-D solid element.

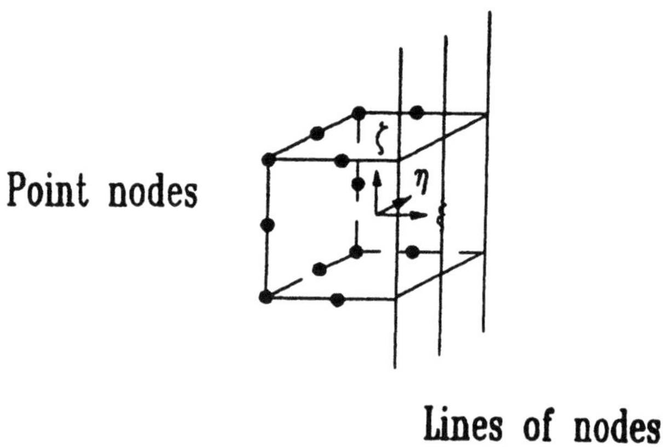

Figure 23 15-node partial hybrid transition element

The Shape Functions of Different Elements

Suppose that the 15-node transition element in Figure 24(b) is used to connect a 20-node solid sub-element (Fig. 24a) to a 8-node plate element (Fig. 24c) in the transition region of a global/local finite element model. On the left side, it meets with a solid element; on the right side, it meets a plate element.

Similar to the case in developing 6-node transition element, take a solid element as an "original" element for developing the transition element. Obviously, the shape functions on the left side do not need to be modified. But, on the right side, new shape functions for satisfying continuity are required.

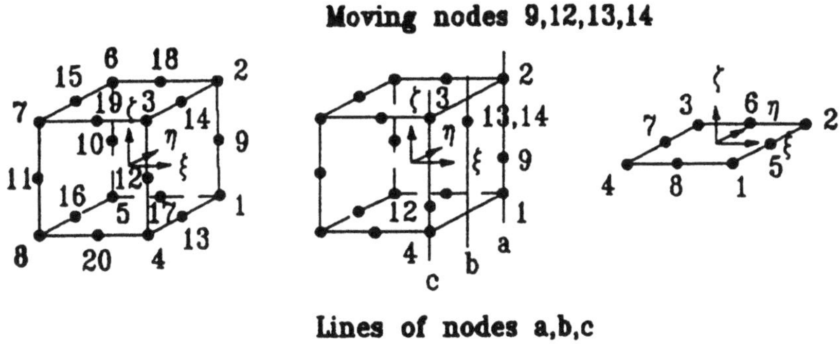

(a) solid element (b) transition element (c) plate element

Figure 24 Three types of element

In Figure 24(a), a typical quadratic solid element is shown with the local curvilinear co-ordinates ξ, η and ζ. Its shape functions can be found elsewhere[4.38-4.39]. For developing a transition element, of particular interest are the shape functions for the nodes on the right side of this element, node 1-4, 9, 12-14:

$$
\begin{aligned}
N_i = & \frac{1}{8}(1+\xi_0)(1+\eta_0)(1+\zeta_0)(\xi_0+\eta_0+\zeta_0-2)\xi_i^2\eta_i^2\zeta_i^2 \\
& +\frac{1}{4}(1-\xi^2)(1+\eta_0)(1+\zeta_0)(1-\xi_i^2)\eta_i^2\zeta_i^2 \\
& +\frac{1}{4}(1-\eta^2)(1+\zeta_0)(1+\xi_0)(1-\eta_i^2)\zeta_i^2\xi_i^2 \\
& +\frac{1}{4}(1-\zeta^2)(1+\xi_0)(1+\eta_0)(1-\zeta_i^2)\xi_i^2\eta_i^2
\end{aligned}
\tag{4-173}
$$

in which,

$$\xi_0 = \xi_i \xi \qquad \eta_0 = \eta_i \eta \qquad \zeta_0 = \zeta_i \zeta \tag{4-174}$$

where ξ_i, η_i and ζ_i are the local co-ordinates of node i in the element parametric space.

The transition element in Fig. 24(b) is similar to the solid element in Fig 24(a) except for the nodes on its right face (1,2,3,4,9,12,13,14). Special treatment has to be done to these nodes so that their displacements can be compatible to those of the plate element in figure 24(c). Consider a function,

$$\Omega(\xi,\eta,\zeta) = \Omega_a(\xi,\eta,\zeta) + \Omega_b(\xi,\eta,\zeta) + \Omega_c(\xi,\eta,\zeta) \tag{4-175}$$

in which,

$$\begin{aligned} \Omega_a(\xi,\eta,\zeta) &= \alpha_1 N_1 + \alpha_2 N_2 + \alpha_9 N_9 \\ \Omega_b(\xi,\eta,\zeta) &= \alpha_{13} N_{13} + \alpha_{14} N_{14} \\ \Omega_c(\xi,\eta,\zeta) &= \alpha_3 N_3 + \alpha_4 N_4 + \alpha_{12} N_{12} \end{aligned} \tag{4-176}$$

where α_i is the value of Ω at the node i of the solid element. Note that the function Ω can have the meaning of displacement function for the right face of the transition element in Fig. 24(b). Functions Ω_a, Ω_b and Ω_c can be of any degree (linear, quadratic etc.) between the two nodes 1-2, 13-14, or 3-4 respectively. For a regular solid element, α_1, α_9 and α_2 would represent the displacements at the node 1, 9 and 2 respectively. Normally, the displacements at node 1, 9 and 2 are independent from each other. However, if the displacements at nodes 1, 9 and 2 are constrained such that each of them is equal to a specific value of a function $\beta(\zeta')$, then one can write:

$$\begin{aligned} \alpha_1 &= \beta(\zeta'_1) \\ \alpha_9 &= \beta(\zeta'_9) \\ \alpha_2 &= \beta(\zeta'_2) \end{aligned} \tag{4 - 177}$$

where ζ'_1, ζ'_9 and ζ'_2 are the global coordinates in the thickness direction of nodes 1, 9 and 2 respectively.

Now consider the shape functions of the degenerated plate element. A middle surface of a degenerated plate element is shown with the local curvilinear co-ordinates ξ and η (in Figure 24(c)). The shape functions for the nodes on the left boundary of this element, nodes 3, 7 and 4 are:

$$\begin{aligned}\overline{N_i}=&\frac{1}{4}(1+\xi_0)(1+\eta_0)(\xi_0+\eta_0-1)\xi_i^2\eta_i^2\\&+\frac{1}{2}(1-\xi^2)(1+\eta_0)(1-\xi_i^2)\eta_i^2\\&+\frac{1}{2}(1-\eta^2)(1+\xi_0)(1-\eta_i^2)\xi_i^2\end{aligned} \tag{4-178}$$

in which,

$$\xi_0=\xi_i\xi \qquad \eta_0=\eta_i\eta \tag{4-179}$$

Consider another function meantime,

$$\Pi(\xi,\eta,\zeta')=\Pi_3(\xi,\eta,\zeta')+\Pi_7(\xi,\eta,\zeta')+\Pi_4(\xi,\eta,\zeta') \tag{4-180}$$

in which

$$\begin{aligned}\Pi_3(\xi,\eta,\zeta')&=A_3\beta_3(\zeta')\overline{N_3}\\\Pi_7(\xi,\eta,\zeta')&=A_7\beta_7(\zeta')\overline{N_7}\\\Pi_4(\xi,\eta,\zeta')&=A_4\beta_4(\zeta')\overline{N_4}\end{aligned} \tag{4-181}$$

Note that the function Π can be considered to be the displacements of points lying on a plane normal to the middle plane of the plate/shell element at edge 3-4. Π_3 can be considered to be the displacements at all points on the normal to the initial mid surface of the undeformed plate element at node 3. Π_7 and Π_4 also can be considered likewise to be the displacements at all points on the normal to the initial middle surface of the undeformed plate element at node 7 and 4 respectively. Let us consider the composition of Π_3 in detail. The compositions of Π_7 and Π_4 follow.

In Π_3 , $\overline{N}_3$ represents the shape function in the plane ξ-η. A_3 represents the nodal displacement at the node 3 on the middle surface of the plate element. $\beta_3(\zeta')$ represents the variation of the displacement in the undeformed state of any point initially lying on the line normal to the middle surface of the plate element. If only one plate element is used for the whole laminate thickness, $\beta(\zeta')$ is a linear function of ζ'.

Matching the Two Shape Functions

From figure 24(b), Ω represents the displacement of the transition element at the interface. From figure 24(c), Π represents the displacement of the plate element at the interface. In order to satisfy the compatibility of displacement fields at the interface between the transition element and plate element, Ω and Π must be the same.

Function Ω consists of three functions Ω_a , Ω_b and Ω_c , and function Π consists of another three functions Π_3, Π_7 and Π_4 . At the interface, Ω_a , Ω_b and Ω_c will need to match Π_3, Π_7 and Π_4 , respectively. If Ω_i and Π_i can be matched exactly, compatibility of displacement fields at the interface will be satisfied. At the interface, ζ' is a function of ζ. In the interval $\zeta \in [-1,1]$ which corresponds to $\zeta' \in [\zeta'_l, \zeta'_{l+1}]$, one has

$$\zeta' = \frac{1-\zeta}{2}\zeta'_l + \frac{1+\zeta}{2}\zeta'_{l+1} \tag{4-140}$$

Note that ζ' is the global thickness coordinate for the plate element while ζ is the smaller thickness coordinate for the solid or transition element.

Without losing generality, the function Ω_a is examined as an example along the line 'a' of nodes 1, 2 and 9. It can be rewritten as follows:

$$\begin{aligned} \Omega_a = {} & \alpha_1 (1+\xi)(1+\eta)(1-\zeta)(\xi+\eta-1)/8 \\ & +\alpha_2 (1+\xi)(1+\eta)(1+\zeta)(\xi+\eta-1)/8 \\ & +\left(\alpha_9 - \frac{\alpha_1+\alpha_2}{2}\right)(1+\xi)(1+\eta)(1-\zeta^2)/4 \end{aligned} \tag{4-182}$$

It shows that the function Ω_a may be separated into two parts: (1) the contribution of the corner nodes which varies linearly along the ζ direction and quadratically along the ξ and η directions; and (2) the contribution of the mid-side node which is quadratic in the ζ direction and linear along ξ and η shown in figure 25(a). If the quadratic function

$$\left(\alpha_9 - \frac{\alpha_1+\alpha_2}{2}\right)(1-\zeta^2) \tag{4-183}$$

can be replaced by a arbitrary function

$$A_3\beta_3(\zeta') - \frac{(1-\zeta)\alpha_1+(1+\zeta)\alpha_2}{2} \tag{4-184}$$

shown in figure 25(b), then Ω_a will be exactly equal to the Π_3.

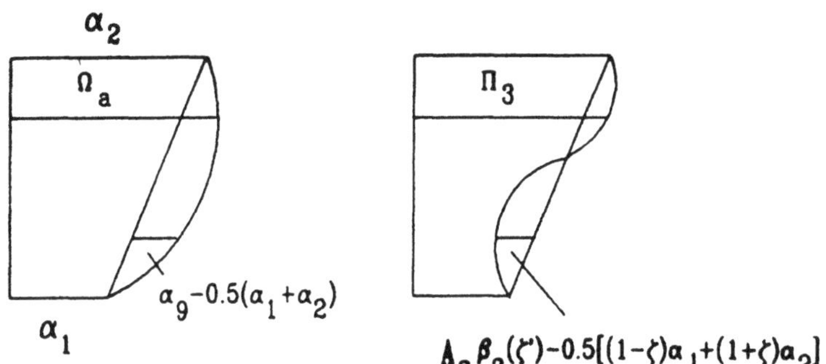

Figure 25 Variation of physical field on the line 'a' of nodes

In order to accommodate any arbitrary curve $\beta_3(\zeta')$ specified by the adjoining plate element, a line of nodes connecting nodes 1-2 and a moving node 9 which moves along this line are defined (see figure 24). At every point ζ' occupied by the moving node, the nodal value α_9 is made to be equal to the value of the specified curve at that point, $A_3\beta_3(\zeta')$ (see equation (4-181). Taking α_1 and α_2 as $A_3\beta_3(\zeta'(-1))$ and $A_3\beta_3(\zeta'(+1))$, respectively, the new shape functions and the nodal value are defined by

$$\begin{aligned} N'_1&=(1+\xi)(1+\eta)(1-\zeta)(\xi+\eta-1)/8\\ N'_2&=(1+\xi)(1+\eta)(1+\zeta)(\xi+\eta-1)/8\\ N'_9&=(1+\xi)(1+\eta)/4\\ \alpha'_9&=A_3\beta_3(\zeta') - \frac{(1-\zeta)\alpha_1+(1+\zeta)\alpha_2}{2} \end{aligned} \tag{4-185}$$

Thus

$$\Omega_a=\alpha_1 N'_1+\alpha_2 N'_2+\alpha'_9 N'_9 \tag{4-186}$$

The equations (4-185) and (4-186) above can be transformed into the most convenient form as follows:

$$\begin{aligned} N''_1 &= (1+\xi)(1+\eta)(1-\zeta)(\xi+\eta-2)/8 \\ N''_2 &= (1+\xi)(1+\eta)(1+\zeta)(\xi+\eta-2)/8 \\ N''_9 &= (1+\xi)(1+\eta)/4 \\ \alpha''_9 &= A_3\beta_3(\zeta') \\ \Omega_a &= \alpha_1 N''_1+\alpha_2 N''_2+\alpha''_9 N''_9 \end{aligned} \tag{4-187}$$

and

$$\Omega_a=A_3[\beta_3(\zeta'_I)N''_1(\xi,\eta,\zeta)+\beta_3(\zeta'_{I+1})N''_2(\xi,\eta,\zeta)+\beta_3(\zeta')N''_9(\xi,\eta)] \tag{4-188}$$

Thus, the Ω_a and Π_3 are matched exactly at the interface. Note that the nodes 1, 2 and 9 are not independent nodes. They become sub-nodes on the line of node 'a'. Because the contribution of line 1-9-2 to the displacement field of the transition element is represented by the function Ω_a, this line is called as line of nodes "a".

In the same way as the functions Ω_a and Π_3, the Ω_b and Ω_c can also be converted to match Π_7 and Π_4. The new shape functions and nodal values are defined by

$$\begin{aligned} N''_{13} &= N''_{14} = (1+\xi)(1-\eta^2)/4 \\ \alpha_{13} &= \alpha_{14} = A_7\beta_7(\zeta') \\ \Omega_b &= \alpha_{13}N''_{13}+\alpha_{14}N''_{14} \end{aligned} \tag{4-189}$$

and

$$\Omega_b=A_7[\beta_7(\zeta')N''_{13}(\xi,\eta)+\beta_7(\zeta')N''_{14}(\xi,\eta)] \tag{4-190}$$

and

$$N''_3=(1+\xi)(1-\eta)(1+\zeta)(\xi-\eta-2)/8$$
$$N''_4=(1+\xi)(1-\eta)(1-\zeta)(\xi-\eta-2)/8$$
$$N''_{12}=(1+\xi)(1-\eta)/4 \qquad (4\text{-}191)$$

$$\alpha''_{12}=A_4\beta_4(\zeta')$$

$$\Omega_c=\alpha_3 N''_3+\alpha_4 N''_4+\alpha''_{12}N''_{12}$$

and

$$\Omega_c=A_4[\beta_4(\zeta'_I)N''_4(\xi,\eta,\zeta)+\beta_4(\zeta'_{I+1})N''_3(\xi,\eta,\zeta)+\beta_4(\zeta')N''_{12}(\xi,\eta)]$$

(4-192)

The functions Ω_a , Ω_b and Ω_c of three lines of nodes "a", "b"and "c" determine the displacements of the transition element at interface. Similarly, the functions Π_3, Π_7 and Π_4 of three nodes 3, 7 and 4 determine the displacements of the plate element at the interface. Because the functions Ω_a, Ω_b and Ω_c are the same as the functions Π_3, Π_7 and Π_4 at the interface respectively, the function Ω is subsequently same as the function Π. Therefore, the displacements are compatible at the interface between the transition element and plate element.

These eight new shape functions (4-187), (4-189) and (4-191) along with the other twelve shape functions as given in Reference [4.38-4.39] form a complete set of shape functions for the transition element.

Geometry of the Element

Renumbering the nodes, a transition element is shown in Figure 26. The global co-ordinate (x, y, z) of any point in the element may be related to the non-dimensional co-ordinates by

$$\begin{Bmatrix}x\\y\\z\end{Bmatrix}=\sum_{1}^{12}N_i\begin{Bmatrix}x_i\\y_i\\z_i\end{Bmatrix}+\sum_{13,15}^{18,20}N''_i\begin{Bmatrix}\overline{x_i}\\\overline{y_i}\\\overline{z_i}\end{Bmatrix}+\sum_{14,16}^{17,19}N''_i\begin{Bmatrix}\overline{x_i}\\\overline{y_i}\\\overline{z_i}\end{Bmatrix}_{moving} \qquad (4\text{-}193)$$

in which, the expression of $(\bar{x}, \bar{y}, \bar{z})$ is dependent on the assumptions used in the adjoining plate element.

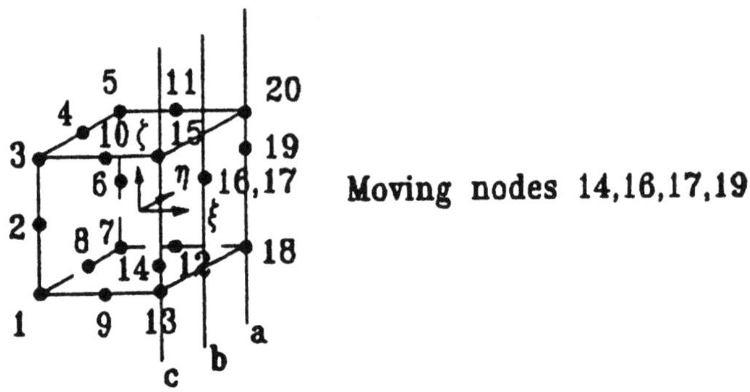

Figure 26 Transition element

When the adjoining element is a degenerated plate element, the equation is written as

$$\begin{Bmatrix} x \\ y \\ z \end{Bmatrix} = \sum_{1}^{12} N_i \begin{Bmatrix} x_i \\ y_i \\ z_i \end{Bmatrix} + \sum_{13,15}^{18,20} N_i \left[\begin{Bmatrix} x_i^0 \\ y_i^0 \\ z_i^0 \end{Bmatrix} + \frac{\zeta'_i}{2} \begin{Bmatrix} \Delta x_i^0 \\ \Delta y_i^0 \\ \Delta z_i^0 \end{Bmatrix} \right] + \sum_{14,16}^{17,19} N_i \left[\begin{Bmatrix} x_i^0 \\ y_i^0 \\ z_i^0 \end{Bmatrix} + \frac{\zeta'}{2} \begin{Bmatrix} \Delta x_i^0 \\ \Delta y_i^0 \\ \Delta z_i^0 \end{Bmatrix} \right] \tag{4-194}$$

in which, x_i, y_i and z_i (i=1,2,..12) are the co-ordinates of node i. x_i^0, y_i^0 and z_i^0 (i=13,14,..20) are the co-ordinates of the lines of node "a", "b", and "c" at the middle surface of the composite structure. The expression (4-194) can be rewritten as follows,

$$\begin{Bmatrix} x \\ y \\ z \end{Bmatrix} = \sum_{1}^{12} N_i \begin{Bmatrix} x_i \\ y_i \\ z_i \end{Bmatrix} + (N''_{18} + N''_{19} + N''_{20}) \begin{Bmatrix} x_a^0 \\ y_a^0 \\ z_a^0 \end{Bmatrix} + \frac{\zeta'_{18} N''_{18} + \zeta' N''_{19} + \zeta'_{20} N''_{20}}{2} \begin{Bmatrix} \Delta x_a^0 \\ \Delta y_a^0 \\ \Delta z_a^0 \end{Bmatrix}$$

$$+ (N''_{16} + N''_{17}) \left(\begin{Bmatrix} x_b^0 \\ y_b^0 \\ z_b^0 \end{Bmatrix} + \frac{\zeta'}{2} \begin{Bmatrix} \Delta x_b^0 \\ \Delta y_b^0 \\ \Delta z_b^0 \end{Bmatrix} \right)$$

$$+(N''_{13}+N''_{14}+N''_{15})\begin{Bmatrix}x_c^0\\y_c^0\\z_c^0\end{Bmatrix}+\frac{\zeta'_{13}N''_{13}+\zeta'N''_{14}+\zeta'_{15}N''_{15}}{2}\begin{Bmatrix}\Delta x_c^0\\\Delta y_c^0\\\Delta z_c^0\end{Bmatrix} \tag{4-195}$$

In the simplified form,

$$\begin{Bmatrix}x\\y\\z\end{Bmatrix}=\sum_1^{12}N_i\begin{Bmatrix}x_i\\y_i\\z_i\end{Bmatrix}+\sum_{a,b}^{c}(N_j\begin{Bmatrix}x_j^0\\y_j^0\\z_j^0\end{Bmatrix}+\frac{\overline{N_j}}{2}\begin{Bmatrix}\Delta x_j^0\\\Delta y_j^0\\\Delta z_j^0\end{Bmatrix}) \tag{4-196}$$

Note that coordinate ζ' is the global thickness coordinate for the plate/shell element and coordinate ζ is the smaller thickness coordinate for the transition element. Note that the co-ordinate ζ' of the moving node varies along the line as follow

$$\zeta'=\frac{1-\zeta}{2}\zeta'_l+\frac{1+\zeta}{2}\zeta'_{l+1} \tag{4-140}$$

where the values ζ'_l and ζ'_{l+1} represent the values of global co-ordinate ζ' at the lower and upper surfaces of a layer while ζ=-1 and ζ=+1, respectively. Denote

$$N_a=N''_{18}+N''_{19}+N''_{20}$$

then

$$\overline{N_a}=\zeta'_{18}N''_{18}+\zeta'N''_{19}+\zeta'_{20}N''_{20}=N_a\zeta'$$

Also, one can obtain

$$N_b=N''_{16}+N''_{17}$$

$$\overline{N_b}=\zeta'(N''_{16}+N''_{17})=N_b\zeta'$$

$$N_c=N''_{13}+N''_{14}+N''_{15}$$

$$\overline{N_c}=\zeta'_{13}N''_{13}+\zeta' N''_{14}+\zeta'_{15}N''_{15}=N_c\zeta' \tag{4-197}$$

Thus, the co-ordinates can be expressed as in simple form,

$$\begin{Bmatrix} x \\ y \\ z \end{Bmatrix}=\sum_{1}^{12} N_i \begin{Bmatrix} x_i \\ y_i \\ z_i \end{Bmatrix}+\sum_{a,b}^{c} N_j \left(\begin{Bmatrix} x_j^0 \\ y_j^0 \\ z_j^0 \end{Bmatrix}+\frac{\zeta'}{2}\begin{Bmatrix} \Delta x_j^0 \\ \Delta y_j^0 \\ \Delta z_j^0 \end{Bmatrix}\right) \tag{4-198}$$

where N_i is the shape function which can be expressed as follows:

$$\begin{aligned} N_i=&\frac{1}{8}(1+\xi_0)(1+\eta_0)(1+\zeta_0)(\xi_0+\eta_0+\zeta_0-2)\xi_i^2\eta_i^2\zeta_i^2 \\ &+\frac{1}{4}(1-\xi^2)(1+\eta_0)(1+\zeta_0)(1-\xi_i^2)\eta_i^2\zeta_i^2 \\ &+\frac{1}{4}(1-\eta^2)(1+\zeta_0)(1+\xi_0)(1-\eta_i^2)\zeta_i^2\xi_i^2 \\ &+\frac{1}{4}(1-\zeta^2)(1+\xi_0)(1+\eta_0)(1-\zeta_i^2)\xi_i^2\eta_i^2 \end{aligned} \tag{4-199}$$

in which

$$\xi_0=\xi_i\xi \quad \eta_0=\eta_i\eta \quad \zeta_0=\zeta_i\zeta \qquad i=1-12 \tag{4-200}$$

and

$$\begin{aligned} N_a&=(1+\xi)(1+\eta)(\xi+\eta-1)/4 \\ N_b&=(1+\xi)(1-\eta^2)/2 \\ N_c&=(1+\xi)(1-\eta)(\xi-\eta-1)/4 \end{aligned} \tag{4-201}$$

It can be seen that N_a, N_b and N_c are same as the shape functions used in the degenerated plate element. Furthermore, The expression (4-198) can be rewritten in the form,

$$\begin{Bmatrix} x \\ y \\ z \end{Bmatrix}=\sum_{1}^{12} N_i \begin{Bmatrix} x_i \\ y_i \\ z_i \end{Bmatrix}+\sum_{a,b}^{c} N_j \left(\begin{Bmatrix} x_j^0 \\ y_j^0 \\ z_j^0 \end{Bmatrix}+\frac{\zeta'}{2}h_j \mathbf{V}_{3j}\right) \tag{4-202}$$

and

$$V_{3j}=\begin{Bmatrix} l_{3j} \\ m_{3j} \\ n_{3j} \end{Bmatrix}=\frac{1}{h_j}\left(\begin{Bmatrix} x_j \\ y_j \\ z_j \end{Bmatrix}_T-\begin{Bmatrix} x_j \\ y_j \\ z_j \end{Bmatrix}_B\right) \tag{4-203}$$

and

$$h_j=\sqrt{(x_{jT}-x_{jB})^2+(y_{jT}-y_{jB})^2+(z_{jT}-z_{jB})^2} \tag{4-204}$$

Displacement Field

In the element, the displacements (see figure 27) are expressed as follows:

$$\begin{Bmatrix} U \\ V \\ W \end{Bmatrix}=\sum_{1}^{12} N_i\begin{Bmatrix} u_i \\ v_i \\ w_i \end{Bmatrix}+\sum_{a,b}^{c} N_i\left[\begin{Bmatrix} u_i^0 \\ v_i^0 \\ w_i^0 \end{Bmatrix}+\zeta'[b_i]\begin{Bmatrix} \Psi_{xi} \\ \Psi_{yi} \\ \Psi_{zi} \end{Bmatrix}\right] \tag{4-205}$$

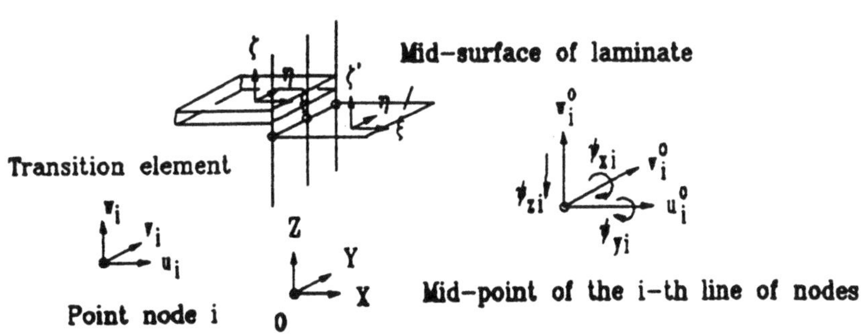

Figure 27 Nodal displacements in a transition element

where

$$[b_i] = \begin{bmatrix} b_{11i} & b_{12i} & b_{13i} \\ b_{21i} & b_{22i} & b_{23i} \\ b_{31i} & b_{32i} & b_{33i} \end{bmatrix} = \frac{h_i}{2} [\mathbf{V_{1i}} \ -\mathbf{V_{2i}} \ \mathbf{V_{3i}}] \tag{4-118}$$

and

$$\mathbf{V_{1i}} = \begin{Bmatrix} l_{1i} \\ m_{1i} \\ n_{1i} \end{Bmatrix} = \frac{\mathbf{i} \times \mathbf{V_{3i}}}{|\mathbf{i} \times \mathbf{V_{3i}}|} \qquad \mathbf{V_{2i}} = \begin{Bmatrix} l_{2i} \\ m_{2i} \\ n_{2i} \end{Bmatrix} = \mathbf{V_{3i}} \times \mathbf{V_{1i}} \tag{4-116}$$

in which, the displacement components u_i , v_i and w_i are the nodal displacements at point nodes 1-12. The components u_i^0, v_i^0 and w_i^0 are the displacements of the line "a", "b" and "c" at the middle surface of the composite structure, ψ_{xi} and ψ_{yi} are two rotations of the nodal vector $\mathbf{V}_{3i}$ about orthogonal directions normal to it, and ψ_{zi} is a transverse normal deformation in the thickness direction.

Partial Strain Field and Partial Derivatives of the Displacement Field

The partial strain field is

$$\{\boldsymbol{\varepsilon}_g\} = \begin{Bmatrix} \frac{\partial u}{\partial x} \\ \frac{\partial v}{\partial y} \\ \frac{\partial u}{\partial y} + \frac{\partial v}{\partial x} \end{Bmatrix} = [B_g]\boldsymbol{\delta} = [B_{g1} B_{g2} \cdot \cdot \cdot \cdot \cdot B_{g15}] \begin{Bmatrix} \boldsymbol{\delta}_1 \\ \boldsymbol{\delta}_2 \\ \cdot \\ \cdot \\ \boldsymbol{\delta}_{15} \end{Bmatrix} \tag{4-206}$$

where $\delta_{13} = \delta_a$, $\delta_{14} = \delta_b$ and $\delta_{15} = \delta_c$. The partial derivatives of the displacement are

$$\mathbf{D_L u} = \begin{Bmatrix} \frac{\partial w}{\partial z} \\ \frac{\partial v}{\partial z} + \frac{\partial w}{\partial y} \\ \frac{\partial w}{\partial x} + \frac{\partial u}{\partial z} \end{Bmatrix} = [B_L]\boldsymbol{\delta} = \{B_{L1} B_{L2} \cdot \cdot \cdot \cdot \cdot B_{L15}] \begin{Bmatrix} \boldsymbol{\delta}_1 \\ \boldsymbol{\delta}_2 \\ \cdot \\ \cdot \\ \boldsymbol{\delta}_{15} \end{Bmatrix} \tag{4-207}$$

For nodes i = 1 - 12,

$$[B_{gi}]=\begin{bmatrix} N_{i,x} & 0 & 0 \\ 0 & N_{i,y} & 0 \\ N_{i,y} & N_{i,x} & 0 \end{bmatrix} \tag{4-208}$$

and

$$[B_{Li}]=\begin{bmatrix} 0 & 0 & N_{i,z} \\ 0 & N_{i,z} & N_{i,y} \\ N_{i,z} & 0 & N_{i,x} \end{bmatrix} \tag{4-209}$$

and

$$\boldsymbol{\delta}_i=\begin{Bmatrix} u_i \\ v_i \\ w_i \end{Bmatrix} \tag{4-210}$$

For nodes i = a , b and c,

$$[B_{gi}]=\begin{bmatrix} N_{i,x} & 0 & 0 & b_{11i}a_{ix} & b_{12i}a_{ix} & b_{13i}a_{ix} \\ 0 & N_{i,y} & 0 & b_{21i}a_{iy} & b_{22i}a_{iy} & b_{23i}a_{iy} \\ N_{i,y} & N_{i,x} & 0 & b_{11i}a_{iy}+b_{21i}a_{ix} & b_{12i}a_{iy}+b_{22i}a_{ix} & b_{13i}a_{iy}+b_{23i}a_{ix} \end{bmatrix} \tag{4-211}$$

and

$$[B_{Li}]=\begin{bmatrix} 0 & 0 & N_{i,z} & b_{31i}a_{iz} & b_{32i}a_{iz} & b_{33i}a_{iz} \\ 0 & N_{i,z} & N_{i,y} & b_{21i}a_{iz}+b_{31i}a_{iy} & b_{22i}a_{iz}+b_{32i}a_{iy} & b_{23i}a_{iz}+b_{33i}a_{iy} \\ N_{i,z} & 0 & N_{i,x} & b_{31i}a_{ix}+b_{11i}a_{iz} & b_{32i}a_{ix}+b_{12i}a_{iz} & b_{33i}a_{ix}+b_{13i}a_{iz} \end{bmatrix}$$

(4-212)

and

(4-213)

$$\boldsymbol{\delta}_i=[u_i^0 \;\; v_i^0 \;\; w_i^0 \;\; \psi_{xi} \;\; \psi_{yi} \;\; \psi_{zi}]^T$$

and

$$
\begin{aligned}
a_{ix} &= N_{i,x}\zeta' + N_i \zeta'_{,x} \\
a_{iy} &= N_{i,y}\zeta' + N_i \zeta'_{,y} \\
a_{iz} &= N_{i,z}\zeta' + N_i \zeta'_{,z}
\end{aligned}
\qquad (4\text{-}214)
$$

In order to calculate $N_{i,x}$, $N_{i,y}$, $N_{i,z}$ and $\zeta'_{,x}$, $\zeta'_{,y}$, $\zeta'_{,z}$, the equation (4-164)-(4-168) for the 6-node transition element are also used.

Assumed Partial Stress Field

In the element, the partial stress field is also assumed independently.

$$
\boldsymbol{\sigma}_g = \begin{Bmatrix} \sigma_z \\ \tau_{yz} \\ \tau_{zx} \end{Bmatrix} = [P_g]\,\boldsymbol{\beta} \qquad (4\text{-}215)
$$

where the stress matrix $[P_g]$ is derived directly from the assumed displacement field using the iso-function method. The iso-function partial stress matrix of the element is

$$
[P_g] = \left| \begin{array}{cccccccccccccccccc}
1 & 0 & 0 & \xi & 0 & 0 & \eta & 0 & 0 & \zeta & 0 & 0 & \xi\eta & 0 & 0 & \xi\zeta & 0 & 0 \\
0 & 1 & 0 & 0 & \xi & 0 & 0 & \eta & 0 & 0 & \zeta & 0 & 0 & \xi\eta & 0 & 0 & \xi\zeta & 0 \\
0 & 0 & 1 & 0 & 0 & \xi & 0 & 0 & \eta & 0 & 0 & \zeta & 0 & 0 & \xi\eta & 0 & 0 & \xi\zeta
\end{array} \right.
$$

$$
\begin{array}{cccccccccccccccc}
\eta\zeta & 0 & 0 & \xi\eta\zeta & 0 & 0 & \xi^2 & 0 & 0 & \eta^2 & 0 & 0 & \zeta^2 & 0 & 0 & \xi^2\eta \\
0 & \eta\zeta & 0 & 0 & \xi\eta\zeta & 0 & 0 & \xi^2 & 0 & 0 & \eta^2 & 0 & 0 & \zeta^2 & 0 & 0 \\
0 & 0 & \eta\zeta & 0 & 0 & \xi\eta\zeta & 0 & 0 & \xi^2 & 0 & 0 & \eta^2 & 0 & 0 & \zeta^2 & 0
\end{array} \qquad (4\text{-}216)
$$

$$
\left. \begin{array}{ccccccccccccc}
0 & 0 & \xi^2\zeta & 0 & \xi\eta^2 & 0 & 0 & \eta^2\zeta & 0 & \xi\zeta^2 & 0 & \eta\zeta^2 & 0 \\
\xi^2\eta & 0 & 0 & \xi^2\zeta & 0 & \xi\eta^2 & 0 & 0 & 0 & 0 & \xi\zeta^2 & 0 & 0 \\
0 & \xi^2\eta & 0 & 0 & 0 & 0 & \xi\eta^2 & 0 & \eta^2\zeta & 0 & 0 & 0 & \eta\zeta^2
\end{array} \right|
$$

There are 47 stress modes in the stress matrix.

Examination of Partial Hybrid Transition Element

In the element, there are twelve point nodes and three lines of nodes. The total number of degrees of freedom equals (n=) 54. The number of the rigid body motions are equal to (r=) 6. Thus, the element has 48 natural deformation modes. Because the rank of the partial stiffness matrix $[K_d]$ equals (n_d=) 30, the matrix $[K_d]$ can represent 30 natural deformation modes of the element. Other 18 natural deformation modes of the element should be represented by the partial stiffness matrix $[K_h]$. Therefore, the assumed stress matrix $[P_g]$ must contain at least 18 stress modes. Based on the iso-function partial stress matrix (4-216), the classification method gives an optimal stress matrix as follows,

$$[P_g] = \begin{bmatrix} 1 & 0 & 0 & \xi & 0 & 0 & \eta & 0 & 0 & \zeta & \xi\eta & 0 & 0 & 0 & \eta\zeta & 0 & 0 & 0 \\ 0 & 1 & 0 & 0 & \xi & 0 & 0 & \eta & 0 & 0 & 0 & \xi\eta & 0 & 0 & 0 & \eta\zeta & \xi\eta\zeta & 0 \\ 0 & 0 & 1 & 0 & 0 & \xi & 0 & 0 & \eta & 0 & 0 & 0 & \xi\eta & \xi\zeta & 0 & 0 & 0 & \xi\eta\zeta \end{bmatrix}$$

(4-217)

The examination of the element shows that there are not any kinematic deformation modes. Three groups of materials are examined. The first is for isotropic material (see Table 15); the second is for anisotropic material (see Table 16); The third is for the composite structure with fibre orientation [90, 0, 90] (see Table 17). In the tables,

$$\lambda_i = \frac{\lambda_{hi}}{\lambda_{ui}} \qquad (4\text{-}218)$$

where λ_{hi} is the eigenvalue of the hybrid element; λ_{ui} is the eigenvalue of the conventional displacement element.

Table 15 Eigenvalue of the Transition Element with 18 stress modes and isotropic materials: E=1100 GPa, ν=0.1

No.	λ_i	No.	λ_i	No.	λ_i
1	0.2071	17	0.8250	33	0.8731
2	0.4803	18	0.7449	34	0.8561
3	0.5921	19	0.8366	35	0.9672
4	0.6132	20	0.8215	36	0.9512
5	0.4749	21	0.8065	37	0.9352
6	0.5130	22	0.7871	38	0.9484
7	0.7002	23	0.8718	39	0.8896
8	0.6885	24	0.9215	40	0.9528
9	0.6687	25	0.7803	41	0.9070
10	0.6114	26	0.7887	42	0.9457
11	0.6351	27	0.8238	43	0.9157
12	0.7420	28	0.8160	44	0.9341
13	0.6695	29	0.8189	45	0.9163
14	0.6840	30	0.8569	46	0.9117
15	0.7370	31	0.7724	47	0.9821
16	0.9066	32	0.7999	48	0.9937

Table 16 Eigenvalue of the Transition Element with 18 stress modes and anisotropic materials: E_L=174.6 GPa, E_T=7.0 GPa, G_{LT}=3.5 GPa, G_{TT}=1.4 GPa, $\nu_{12}=\nu_{13}=\nu_{23}$=0.25

No.	λ_i	No.	λ_i	No.	λ_i
1	0.5415	17	0.8292	33	0.9141
2	0.5238	18	0.9127	34	0.9506
3	0.5200	19	0.8954	35	0.9509
4	0.6377	20	0.7824	36	0.8810
5	0.6938	21	0.8598	37	0.8568
6	0.6149	22	0.8410	38	0.8714
7	0.7558	23	0.9131	39	0.9966
8	0.7858	24	0.8118	40	0.9768
9	0.6984	25	0.9484	41	0.9976
10	0.5825	26	0.8229	42	0.9977
11	0.6447	27	0.8601	43	0.9921
12	0.7881	28	0.8674	44	0.9961
13	0.8234	29	1.0000	45	0.9907
14	0.8774	30	0.9756	46	0.9993
15	0.9520	31	0.9746	47	0.9959
16	0.8757	32	0.9410	48	0.9993

Table 17 Eigenvalue of the Transition Element with fibre orientation [90^0, 0^0, 90^0], 18 stress modes and materials: E_L=174.6 GPa, E_T=7.0 GPa, G_{LT}=3.5 GPa, G_{TT}=1.4 GPa, $\nu_{12}=\nu_{13}=\nu_{23}$=0.25

No.	λ_i	No	λ_i	No.	λ_i
1	0.1828	17	0.7402	33	0.9464
2	0.4224	18	0.8149	34	0.8900
3	0.3981	19	0.7365	35	0.8766
4	0.4885	20	0.8655	36	0.9078
5	0.3759	21	0.7400	37	0.9665
6	0.5307	22	0.7757	38	0.9945
7	0.5726	23	0.7921	39	0.9928
8	0.8220	24	0.8156	40	0.9990
9	0.5805	25	0.7856	41	0.9913
10	0.6928	26	0.8156	42	0.9991
11	0.6623	27	0.8038	43	0.9951
12	0.7036	28	0.9790	44	0.9989
13	0.6086	29	0.9930	45	0.9961
14	0.5536	30	0.9130	46	0.9987
15	0.8042	31	0.9350	47	0.9937
16	0.8358	32	0.8143	48	0.9996

4.3 MULTILAYER FINITE ELEMENTS

The multilayer finite elements have been widely used for stress analysis of composite structures. Usually, a multilayer element consists of a stack of sub-elements. According to the distribution of the material layers, a composite structure is divided into many sub-layers along the thickness and each sub-layer is modeled by a sub-element. When the matrices of sub-elements are formulated, they are assembled through the thickness using continuity conditions at the interfaces between different sub-elements, and then the multilayer element matrices are obtained. Therefore, there are two steps to obtain a multilayer element matrix: the first is to formulate the sub-element matrices and the second is to assemble them to form a multilayer element matrix. In this section, two multilayer elements [4.40-4.41] are presented.

4.3.1 Fomulation of Partial Hybrid Multilayer Element

In section 4.2.1, the composite variational functional has been expressed in the form,

$$\Pi_{co}=\int_V [\frac{1}{2}\boldsymbol{\varepsilon}_g^T[R_1]\,\boldsymbol{\varepsilon}_g+\frac{1}{2}\boldsymbol{\sigma}_g^T[R_3]\,\boldsymbol{\sigma}_g+\boldsymbol{\sigma}_g^T[R_2]^T\boldsymbol{\varepsilon}_g + \boldsymbol{\sigma}_g\boldsymbol{D}_L\mathbf{u}-\boldsymbol{F}^T\mathbf{u}]\,dV-\int_{S_t}\boldsymbol{T}^T\mathbf{u}\,dS \tag{4-5}$$

in which, the layer material matrix [R] is

$$[R]=\begin{bmatrix} R_1 & R_2 \\ R_2^T & R_3 \end{bmatrix}=\begin{bmatrix} S_1^{-1} & -S_1^{-1}S_2 \\ -S_2^TS_1^{-1} & S_2^TS_1^{-1}S_2-S_3 \end{bmatrix} \tag{4-4}$$

or

$$[R]=\begin{bmatrix} R_1 & R_2 \\ R_2^T & R_3 \end{bmatrix}=\begin{bmatrix} C_1-C_2C_3^{-1}C_2^T & C_2C_3^{-1} \\ C_3^{-1}C_2^T & -C_3^{-1} \end{bmatrix} \tag{4-4'}$$

where [S] is the compliance matrix of layer materials and [C] is the stiffness matrix of layer materials. If a composite structure contains N different material layers, the multilayer element will consist of N sub-elements (see figure 28). Therefore, the variational functional becomes

$$\Pi_{co}=\sum_{i=1}^{N}\int_{V}[\frac{1}{2}\boldsymbol{\varepsilon}_g^{iT}[R_1^i]\,\boldsymbol{\varepsilon}_g^i+\frac{1}{2}\boldsymbol{\sigma}_g^{iT}[R_3^i]\,\boldsymbol{\sigma}_g^i+\boldsymbol{\sigma}_g^{iT}[R_2^i]^T\boldsymbol{\varepsilon}_g^i$$
$$+\boldsymbol{\sigma}_g^{iT}\boldsymbol{D}_L\boldsymbol{u}^i-\boldsymbol{F}^T\boldsymbol{u}^i]\,dV-\int_S\boldsymbol{T}^T\boldsymbol{u}^i\,dS \tag{4-219}$$

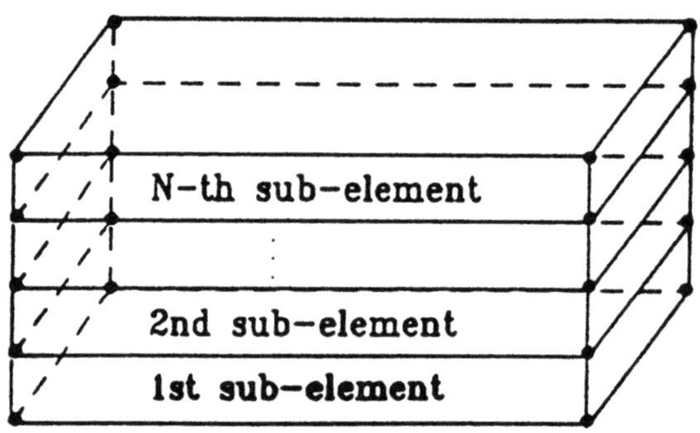

Figure 28 A multilayer element

Sub-Element Matrices

In multilayer element, the displacement field and the partial stress field must be assumed within each sub-element. Suppose the displacement fields in different sub-elements have the same expression form. Similarly, assume that the partial stress fields in different sub-elements have the same expression form. Thus, for the i-th sub-element, the displacement field is assumed as

$$\boldsymbol{u}^i=\begin{Bmatrix} u^i \\ v^i \\ w^i \end{Bmatrix}=[N]\,\boldsymbol{\delta}^i \tag{4-220}$$

where [N] is the matrix of shape functions. Then, the partial strains are

$$\boldsymbol{\varepsilon}_g^i=\begin{Bmatrix}\varepsilon_x^i\\ \varepsilon_y^i\\ \varepsilon_{xy}^i\end{Bmatrix}=\boldsymbol{D}_g\boldsymbol{u}^i=\left[\frac{\partial u^i}{\partial x},\frac{\partial v^i}{\partial y},\frac{\partial u^i}{\partial y}+\frac{\partial v^i}{\partial x}\right]^T=[B_g]\,\boldsymbol{\delta}^i \tag{4-221}$$

and the partial derivatives are

$$\boldsymbol{D}_L\boldsymbol{u}^i=\left[\frac{\partial w^i}{\partial z},\frac{\partial w^i}{\partial y}+\frac{\partial v^i}{\partial z},\frac{\partial w^i}{\partial x}+\frac{\partial u^i}{\partial z}\right]^T=[B_L]\,\boldsymbol{\delta}^i \tag{4-222}$$

in which, $[B_g]$ is a partial geometry matrix and $[B_L]$ is a partial derivative matrix. Along the thickness of the sub-element, a partial stress field is also assumed independently as

$$\boldsymbol{\sigma}_g^i=\begin{Bmatrix}\sigma_z^i\\ \sigma_{yz}^i\\ \sigma_{zx}^i\end{Bmatrix}=[P_g]\,\boldsymbol{\beta}^i=[\boldsymbol{\sigma}_{g1}\ \boldsymbol{\sigma}_{g2}\cdots\boldsymbol{\sigma}_{gl}]\begin{Bmatrix}\beta_1^i\\ \beta_2^i\\ \cdot\\ \cdot\\ \beta_l^i\end{Bmatrix} \tag{4-223}$$

where $[P_g]$ is an assumed stress matrix, $\{\sigma_{gj}\}$ are partial stress modes, and β_j^i are stress parameters. Substituting equations (4-220)-(4-223) into the composite energy functional (4-219), the functional becomes

$$\begin{aligned}\Pi_{co}=\sum_{i=1}^{N}\{&\frac{1}{2}\boldsymbol{\delta}^{iT}\int_{V_i}[B_g]^T[R_1^i]\,[B_g]\,dV\,\boldsymbol{\delta}\\&+\frac{1}{2}\boldsymbol{\beta}^{iT}\int_{V_i}[P_g]^T[R_3^i]\,[P_g]\,dV\,\boldsymbol{\beta}\\&+\boldsymbol{\beta}^{iT}\int_{V_i}[P_g]^T(\,[B_L]+[R_2^i]^T[B_g]\,)\,dV\,\boldsymbol{\delta}\,\}\\&-\boldsymbol{\delta}^{iT}\int_{V_i}[N]^T\boldsymbol{F}dV\ -\boldsymbol{\delta}^{iT}\int_{S_{ti}}[N]^T\boldsymbol{T}dS\end{aligned} \tag{4-224}$$

Denote

$$
\begin{aligned}
&[H^i] = -\int_{V_i} [P_g]^T [R_3^i] [P_g] dV \\
&[G^i] = \int_{V_i} [P_g]^T ([B_L] + [R_2^i]^T [B_g]) dV \\
&[K_d^i] = \int_{V_i} [B_g]^T [R_1^i] [B_g] dV \\
&\boldsymbol{f}^i = \int_{V_i} [N]^T \boldsymbol{F} dV + \int_{S_{ti}} [N]^T \boldsymbol{T} dS
\end{aligned}
\tag{4-225}
$$

The matrices in the equation (4-225) are the sub-element matrices. They will be assembled using continuity conditions at the interfaces between different sub-elements.

Multilayer Matrices

Using the expression (4-225), the variational functional (4-224) can be expressed as

$$
\begin{aligned}
\Pi_{co} = \sum_{i=1}^{N} (& \frac{1}{2} \boldsymbol{\delta}^{iT} [K_d^i] \boldsymbol{\delta}^i - \frac{1}{2} \boldsymbol{\beta}^{iT} [H^i] \boldsymbol{\beta}^i \\
& + \boldsymbol{\beta}^{iT} [G^i] \boldsymbol{\delta}^i - \boldsymbol{\delta}^{iT} \boldsymbol{f}^i)
\end{aligned}
\tag{4-226}
$$

In this variation functional, the stress parameters are not independent and they must be replaced by independent parameters using continuity conditions at interfaces between different sub-elements. There are three ways to formulate multilayer matrices.

i) Laminate Stress Parameters

In general, the stress parameters β_j^i in equation (4-223) are internal parameters, called layer stress parameters. They are not independent and the sub-element matrices can not be assembled based on these layer stress parameters. The constraint of interlaminar surface traction continuity must be used. This constraint requires that σ_{xz}, σ_{yz} and σ_z be continuous at interlaminar surfaces. Therefore, stresses at the lower surface (ζ=-1) of sub-element i+1 must be equal to those at the upper surface (ζ=+1) of sub-element i as follows:

$$\begin{aligned}\sigma_{xz}^{i+1}\big|_{\zeta=-1} &= \sigma_{xz}^{i}\big|_{\zeta=+1}\\ \sigma_{yz}^{i+1}\big|_{\zeta=-1} &= \sigma_{yz}^{i}\big|_{\zeta=+1}\\ \sigma_{z}^{i+1}\big|_{\zeta=-1} &= \sigma_{z}^{i}\big|_{\zeta=+1}\end{aligned} \tag{4-227}$$

Substituting the assumed partial stress field (4-223) into the continuity condition (4-227), the relationships between β_j^i and β_j^{i-1} can be obtained and some dependent stress parameters can be eliminated. Suppose the parameter vector β contains m independent stress parameters, called laminate stress parameters [4.42-4.43]. Thus, the internal layer stress parameters β^i can be replaced by the laminate stress parameter β. Corresponding to the interface continuity conditions (4-227), internal layer stress parameters β^i and nodal displacements δ^i can be related to the laminate stress parameters β and nodal displacements δ , respectively, in the form;

$$\begin{aligned}\boldsymbol{\beta}^i &= [C]_s^i\,\boldsymbol{\beta}\\ \boldsymbol{\delta}^i &= [C]_d^i\,\boldsymbol{\delta}\end{aligned} \tag{4-228}$$

where $[C]^i_s$ and $[C]^i_d$ are assembling matrices. Therefore, the summation over the layers can be taken inside and the multilayer element matrix defined as:

$$\begin{aligned}[K_d] &= \sum_{i=1}^{N} [C]_d^{iT}[K_d^i]\,[C]_d^i\\ [H] &= \sum_{i=1}^{N} [C]_s^{iT}[H^i]\,[C]_s^i\\ [G] &= \sum_{i=1}^{N} [C]_s^{iT}[G^i]\,[C]_d^i\\ \boldsymbol{f} &= \sum_{i=1}^{N} [C]_d^{iT}\boldsymbol{f}^i\end{aligned} \tag{4-229}$$

These operations are analogous to element 'assembly' operations; a set of layer-to-laminate stress parameter 'pointers' and nodal displacement 'pointers' can be used to locate (and add in) sub-element matrix contributions in the multilayer element matrices. Now the variational functional (4-226) becomes

$$\Pi_{co}=\frac{1}{2}\boldsymbol{\delta}^T[K_d]\,\boldsymbol{\delta}-\frac{1}{2}\boldsymbol{\beta}^T[H]\,\boldsymbol{\beta}+\boldsymbol{\beta}^T[G]\,\boldsymbol{\delta}-\boldsymbol{\delta}^T\boldsymbol{f} \tag{4-230}$$

ii) Internal Stress Parameter and Surface Stress parameter

In the equation (4-228), the internal layer stress parameters β^i are transferred to laminate stress parameters β. On the other hand, they also can be transferred to another kind of stress parameters α, called surface stress parameters [4.20,4.40]. The continuous conditions are expressed in the form

$$\boldsymbol{\beta}^i=[U]\begin{Bmatrix}\boldsymbol{\alpha}^i\\ \boldsymbol{\alpha}^{i+1}\end{Bmatrix} \tag{4-231}$$

where α^i is related to the lower surface of the i-th sub-element and α^{i+1} to the upper surface. The matrix [U] transfers internal stress parameters β^i (i=1,2, ... N) to surface stress parameters α^i (i=1,2, ... N+1). The assumed stress field is expressed in terms of surface stress parameters,

$$\boldsymbol{\sigma}_g^i=[P]_g[U]\begin{Bmatrix}\boldsymbol{\alpha}^i\\ \boldsymbol{\alpha}^{i+1}\end{Bmatrix} \tag{4-232}$$

For convenience, it is rewritten as

$$\boldsymbol{\sigma}_g^i=[\overline{P}]_g\,\overline{\boldsymbol{\alpha}}^i \tag{4-233}$$

where

$$[\overline{P}]_g=[P]_g[U] \qquad \overline{\boldsymbol{\alpha}}^i=\begin{Bmatrix}\boldsymbol{\alpha}^i\\ \boldsymbol{\alpha}^{i+1}\end{Bmatrix} \tag{4-234}$$

Thus, the matrices $[H^i]$ and $[G^i]$ in the equation (4-225) become

$$[H^i]' = -\int_{V_i} [\overline{P_g}]^T [R_3^i] [\overline{P_g}] \, dV$$
$$[G^i]' = \int_{V_i} [\overline{P_g}]^T ([B_L] + [R_2^i]^T [B_g]) \, dV \qquad (4\text{-}235)$$

Then, the variational functional takes the form

$$\Pi_{co} = \sum_{i=1}^{N} \left(\frac{1}{2} \boldsymbol{\delta}^{iT} [K_d^i] \boldsymbol{\delta}^i - \frac{1}{2} \overline{\boldsymbol{\alpha}}^{iT} [H^i]' \overline{\boldsymbol{\alpha}}^i + \overline{\boldsymbol{\alpha}}^{iT} [G^i]' \boldsymbol{\delta}^i - \boldsymbol{\delta}^{iT} \boldsymbol{f}^i \right) \qquad (4\text{-}236)$$

In order to assemble all the sub-elements in the multilayer element from 1 to N, define the assembling rule as

$$\boldsymbol{\delta} = \sum_{i=1}^{N} \boldsymbol{\delta}^i = [\boldsymbol{d}^1 \ \boldsymbol{d}^2 \ \ldots \ \boldsymbol{d}^{N+1}]^T \qquad (4\text{-}237)$$

where $\mathbf{d}^k$ is the nodal displacement vector at the k-th surface, and

$$\boldsymbol{\alpha} = \sum_{i=1}^{N} \boldsymbol{\alpha}^i = [\boldsymbol{\alpha}^1 \ \boldsymbol{\alpha}^2 \ \ldots \ \boldsymbol{\alpha}^{N+1}]^T \qquad (4\text{-}238)$$

Applying these assembling rules, the multilayer element matrices are obtained by

$$[K]_d = \sum_{i=1}^{N} [K_d^i] \qquad [H] = \sum_{i=1}^{N} [H^i]'$$
$$[G] = \sum_{i=1}^{N} [G^i]' \qquad \boldsymbol{f} = \sum_{i=1}^{N} \boldsymbol{f}^i \qquad (4\text{-}239)$$

Now, the variational functional (4-226) becomes

$$\Pi_{co} = \frac{1}{2} \boldsymbol{\delta}^T [K_d] \boldsymbol{\delta} - \frac{1}{2} \boldsymbol{\alpha}^T [H] \boldsymbol{\alpha} + \boldsymbol{\alpha}^T [G] \boldsymbol{\delta} - \boldsymbol{\delta}^T \boldsymbol{f} \qquad (4\text{-}230)'$$

iii) Surface Stress Parameter

In the two approaches above, it is necessary to transfer internal stress parameters (or layer stress parameters) to surface stress parameters (or laminate stress parameters). However, if surface stress parameters are used directly in assumed partial stress field (4-223), the transformation will be not necessary. One can assume a partial stress field in the following form,

$$\boldsymbol{\sigma}_g^i = [P_g]\,\boldsymbol{\phi}^i = [P]\,\frac{1}{2}\left\{(1+\zeta)\,\boldsymbol{\alpha}_T^i + (1-\zeta)\,\boldsymbol{\alpha}_B^i\right\} \tag{4-240}$$

where α_T^i and α_B^i are the surface stress parameters corresponding to upper and lower surfaces of the i-th sub-element, respectively. In this expression, a stress mode σ_j in the matrix [P] is related to both of upper and lower surfaces α_T^i and α_B^i and corresponds two stress modes $0.5*(1+\zeta)*\sigma_j$ and $0.5*(1-\zeta)*\sigma_j$ in the assumed stress matrix $[P_g]$. The matrix [P] is determined by displacement polynomials, iso-function method, and classification method. At the interface between the sub-element i and i+1, the surface stress parameters α_T^i is the same as α_B^{i+1}. This means:

$$\boldsymbol{\alpha}_T^i = \boldsymbol{\alpha}_B^{i+1} \tag{4-241}$$

Furthermore, the continuity condition at interface of the laminated structure can be expressed as:

$$\boldsymbol{\sigma}_g^i\big|_{\zeta^i=1} = \boldsymbol{\sigma}_g^{i+1}\big|_{\zeta^{i+1}=-1} \tag{4-242}$$

Thus, one can obtain the condition

$$[P]\,\big|_{\zeta^i=+1} = [P]\,\big|_{\zeta^{1+1}=-1} \tag{4-243}$$

Therefore, the matrix [P] must be a function consisted of even order terms of the coordinate ζ. In order to assemble all the sub-elements in the multilayer element, define the assembling rule as

$$\boldsymbol{\delta} = \sum_{i=1}^{N} \boldsymbol{\delta}^i = [\boldsymbol{d}^1\ \ \boldsymbol{d}^2\ \ \ldots\ \ \boldsymbol{d}^{N+1}]^T \tag{4-237}$$

where $\mathbf{d}^k$ is the nodal displacement vector at the k-th surface, and

$$\boldsymbol{\varphi}=\sum_{i=1}^{N}\boldsymbol{\alpha}^{i}=[\boldsymbol{\alpha}^{1}\ \boldsymbol{\alpha}^{2}\ \ldots\ \boldsymbol{\alpha}^{N+1}]^{T} \tag{4-244}$$

Applying these assembling rules, the multilayer element matrices are obtained by

$$[K]_{d}=\sum_{i=1}^{N}[K_{d}^{i}] \qquad [H]=\sum_{i=1}^{N}[H^{i}]$$
$$[G]=\sum_{i=1}^{N}[G^{i}] \qquad \mathbf{f}=\sum_{i=1}^{N}\mathbf{f}^{i} \tag{4-239}$$

Now, the variational functional (4-226) becomes

$$\Pi_{co}=\frac{1}{2}\boldsymbol{\delta}^{T}[K_{d}]\,\boldsymbol{\delta}-\frac{1}{2}\boldsymbol{\varphi}^{T}[H]\,\boldsymbol{\varphi}+\boldsymbol{\varphi}^{T}[G]\,\boldsymbol{\delta}-\boldsymbol{\delta}^{T}\mathbf{f} \tag{4-245}$$

Using three different approaches, the final expression form of the variational functional (4-230), (4-230)', and (4-245) are the same. After obtaining multilayer element matrices, the variational functional can be written in general form,

$$\Pi_{co}=\frac{1}{2}\boldsymbol{\delta}^{T}[K_{d}]\,\boldsymbol{\delta}-\frac{1}{2}\boldsymbol{\beta}^{T}[H]\,\boldsymbol{\beta}+\boldsymbol{\beta}^{T}[G]\,\boldsymbol{\delta}-\boldsymbol{\delta}^{T}\mathbf{f} \tag{4-12}$$

Then, similar to single-layer element, the stiffness matrix of the multilayer element can be derived using the variational principle of composite energy,

$$[K]=[K_{d}]+[K_{h}]$$
$$[K_{h}]=[G]^{T}[H]^{-1}[G] \tag{4-17}$$

in which, [K] is the element stiffness matrix, the semi-stiffness matrix $[K_d]$ is a displacement-formulated stiffness matrix based on the globally continuous strains, and the semi-stiffness matrix $[K_h]$ is a hybrid-formulated stiffness matrix based on the globally continuous stresses. Then, the governing equation of the multilayer element is

$$[K]\,\boldsymbol{\delta}=\boldsymbol{f} \tag{4-18}$$

After obtaining the nodal displacement δ by means of system equations, the displacement field, stress field, and strain field can be obtained using the following equations:

1. Displacement field

$$\boldsymbol{u}^{i}=\begin{Bmatrix} U^{i} \\ V^{i} \\ W^{i} \end{Bmatrix}=[N]\,\boldsymbol{\delta}^{i} \tag{4-246}$$

2. Partial globally continuous strains

$$\boldsymbol{\varepsilon}_{g}^{i}=\begin{Bmatrix} \varepsilon_{x}^{i} \\ \varepsilon_{y}^{i} \\ \varepsilon_{xy}^{i} \end{Bmatrix}=\boldsymbol{D}_{g}\boldsymbol{u}^{i}=[B_{g}]\,\boldsymbol{\delta}^{i} \tag{4-247}$$

3. Partial globally continuous stresses

$$\boldsymbol{\beta}=[H]^{-1}[G]\,\boldsymbol{\delta} \tag{4-14}$$

$$\boldsymbol{\sigma}_{g}^{i}=\begin{Bmatrix} \sigma_{z}^{i} \\ \sigma_{yz}^{i} \\ \sigma_{zx}^{i} \end{Bmatrix}=[P_{g}]\,\boldsymbol{\beta}^{i} \tag{4-248}$$

4. Partial locally continuous stresses within i-th layer

$$\boldsymbol{\sigma}_{L}^{i}=\begin{Bmatrix} \sigma_{x}^{i} \\ \sigma_{y}^{i} \\ \sigma_{xy}^{i} \end{Bmatrix}=[R_{1}^{i}]\,\boldsymbol{\varepsilon}_{g}^{i}+[R_{2}^{i}]\,\boldsymbol{\sigma}_{g}^{i} \tag{4-249}$$

5. Partial locally continuous strains within i-th layer

$$\boldsymbol{\varepsilon}_L^i = \begin{Bmatrix} \varepsilon_z^i \\ \varepsilon_{yz}^i \\ \varepsilon_{zx}^i \end{Bmatrix} = -[R_2^i]^T \boldsymbol{\varepsilon}_g^i - [R_3^i]\, \boldsymbol{\sigma}_g^i \tag{4-250}$$

4.3.2 Multilayer Solid Element

Partial hybrid multilayer elements consist of a stack of partial hybrid sub-elements. So the elements formulated by the variational principle of composite energy can be used as sub-elements. For examples, 3-D, 8-node partial hybrid solid element and 3-D, 20-node partial hybrid solid element can be used to construct partial hybrid multilayer solid elements. For simplicity, a multilayer element based on 3-D, 8-node solid elements is presented.

Sub-Element Matrices

The multilayer solid element consists of a stack of 3-D, 8-node solid elements (see figure 29). For the i-th sub-element, the assumed displacement field is the same as that for 3-D, 8-node solid element in section 4.2.2. It is in the form,

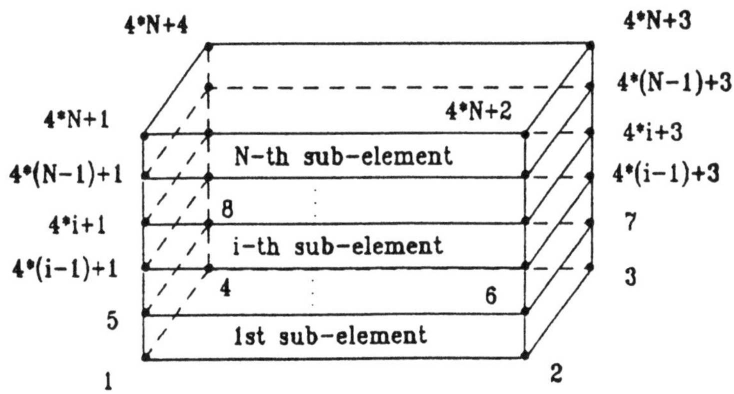

Figure 29 Multilayer solid element

$$u^i = [N]\,\delta^i = [N_1 I \;\; N_2 I \;\; \ldots \;\; N_8 I]\begin{Bmatrix} \delta_1^i \\ \delta_2^i \\ \cdots \\ \delta_8^i \end{Bmatrix} \tag{4-251}$$

in which, [I] is a 3×3 unit matrix, N_i is the shape functions, and the nodal displacement vector is

$$\delta_j^i = \begin{Bmatrix} u_j^i \\ v_j^i \\ w_j^i \end{Bmatrix} \qquad j=1,2,\ldots,8 \tag{4-252}$$

The nodal dispalcement vector of the i-th sub-element can be written in another form,

$$\begin{aligned} \delta^i &= [d^i \;\; d^{i+1}] \\ &= [d_1^i \;\; d_2^i \;\; d_3^i \;\; d_4^i \;\; d_1^{i+1} \;\; d_2^{i+1} \;\; d_3^{i+1} \;\; d_4^{i+1}] \end{aligned} \tag{4-253}$$

where d^i is the nodal displacement vectors related to the lower surface of the sub-element, and d^{i+1} is the nodal displacement vectors related to the upper surface of the sub-element. Within the sub-element, the the partial stress field is assumed in the form

$$\sigma_g^i = \begin{Bmatrix} \sigma_z^i \\ \sigma_{yz}^i \\ \sigma_{zx}^i \end{Bmatrix} = [P_g]\,\phi^i = [P]\,\frac{1}{2}\{(1+\zeta)\,\alpha_T^i + (1-\zeta)\,\alpha_B^i\} \tag{4-254}$$

where α_T^i and α_B^i are the surface stress parameters corresponding to upper and lower surfaces of the i-th sub-element, respectively. When the matrix [P] is a function consisted of even order terms of the coordinate ζ, the continuity condition at interfaces will be automatically satisfied. Using the equations (4-11) and (4-21)-(4-41), the sub-element matrices can be obtained as follows,

$$[H^i] = -\int_{V_i} [P_g]^T [R_3^i] [P_g] dV$$
$$[G^i] = \int_{V_i} [P_g]^T ([B_L] + [R_2^i]^T [B_g]) dV$$
$$[K_d^i] = \int_{V_i} [B_g]^T [R_1^i] [B_g] dV \tag{4-225}$$
$$\mathbf{f}^i = \int_{V_i} [N]^T \mathbf{F} dV + \int_{S_{ti}} [N]^T \mathbf{T} dS$$

Multilayer Matrices

Applying the assembling rules (4-237) and (4-244),

$$\boldsymbol{\delta} = \sum_{i=1}^{N} \boldsymbol{\delta}^i = [\mathbf{d}^1 \ \mathbf{d}^2 \ \ldots \ \mathbf{d}^{N+1}]^T \tag{4-237}$$

and

$$\boldsymbol{\varphi} = \sum_{i=1}^{N} \boldsymbol{\alpha}^i = [\boldsymbol{\alpha}^1 \ \boldsymbol{\alpha}^2 \ \ldots \ \boldsymbol{\alpha}^{N+1}]^T \tag{4-244}$$

the sub-element matrices from the 1st layer to N-th layer are added to form the multilayer matrices,

$$[K]_d = \sum_{i=1}^{N} [K_d^i] \qquad [H] = \sum_{i=1}^{N} [H^i]$$
$$[G] = \sum_{i=1}^{N} [G^i] \qquad \mathbf{f} = \sum_{i=1}^{N} \mathbf{f}^i \tag{4-239}$$

Then, the stiffness matrix of the multilayer element can be calculated using equation (4-17),

$$[K] = [K_d] + [K_h]$$
$$[K_h] = [G]^T [H]^{-1} [G] \tag{4-17}$$

Examination of the Element

For a single-layer element, a necessary and sufficient condition for guaranteeing the absence of kinematic deformation modes at the element level is,

$$n_h = n - r - n_d \tag{4-44}$$

But, for a multilayer element, the minimum number of stress modes in an assumed stress matrix varies with the number of sub-elements in the multilayer elements. Using eigenvalue examination of matrices, the rank n_d of the displacement-formulated stiffness matrix $[K_d]$ can be calculated for different multilayer elements with different number of sub-elements. The minimum number n_h of stress modes in an assumed partial stress matrix is given in table 18. In the table, N is the total number of sub-elements in the multilayer element; n is the total degrees of freedom of the multilayer element; r is the number of rigid body motions.

Table 18 Minimum number of stress modes in the matrix $[P_g]$

N	n	r	n_d	n_h
1	24	6	10	8
2	36	6	15	15
3	48	6	20	22
.	.	.	.	.
10	132	6	55	71
.	.	.	.	.

From table 18, it is observed that the rank of semi-stiffness matrix $[K_d]$

increases by 5 when the multilayer element increases a surface. For example, there are two surfaces in a fundamental multilayer element consisted of one sub-element, and the rank of semi-stiffness matrix equals 10. There is one increased surface in the multilayer element consisted of two sub-elements, and the rank of matrix $[K_d]$ equals 15. Furthermore, there are eleven surfaces in the multilayer element consisted of ten sub-elements, and the rank of matrix $[K_d]$ is equal to 55. Thus, each increased surface in a multilayer element corresponds to 5 deformation modes related to semi-stiffness matrix $[K_d]$. Meanwhile, each added surface will increase four point nodes which correspond 12 degrees of freedom in the multilayer element. Thus, if a multilayer element contains N layers, it will have N+1 surfaces and one has

$$n = 12 * (N+1) \qquad n_d = 5 * (N+1) \quad r = 6 \tag{4-255}$$

Define that m' is the number of stress modes in matrix [P] related to a surface. Thus, the total number of stress modes for the multilayer element is

$$m = (N+1)\ m' \tag{4-256}$$

The necessary and sufficient condition (4-44) for avoiding kinematic deformation modes is

$$(N+1)\ m' = 12*(N+1) - 5*(N+1) - 6 \tag{4-257}$$

Therefore, one obtains minimum number of stress modes in the matrix [P] related to each surface for the multilayer element as follows,

$$m' = 7 - \frac{6}{N+1} \tag{4-258}$$

Using this formulation, one can calculate the number of stress modes related to each surface for multilayer element consisted of different number of sub-elements.

N = 1	m' = 4
N = 2	m' = 5
N = 3	m' = 6
N = 4	m' = 6
N = 5	m' = 6
N = 6	m' = 7
....	....
N = 100	m' = 7

Therefore, the number of stress modes needed in the stress matrix [P] in order to avoid kinematic deformation modes is different for different N.

Sub-Element Stiffness Matrix and Kinematic Deformation Modes

When N=1, the multilayer element becomes a sub-element. The sub-element has (n=) 24 degrees of freedom and (r=) 6 degrees of the rigid displacement. Thus, the sub-element has 18 natural deformation modes. The eigenvalue examination indicates that the rank of the partial stiffness matrix $[K_d]$ for the sub-element is $(n_d=)$ 10. Therefore, the partial stiffness matrix $[K_d]$ represents 10 natural deformation modes of the element, and another partial stiffness matrix $[K_h]$ represents 8 natural deformation modes. Therefore, the minimum number of the stress modes in the assumed stress field $[P_g]$ is equal to 8. Since a stress mode σ_j in the stress matrix [P] represents two stress modes $0.5*(1+\zeta)*\sigma_j$ and $0.5*(1-\zeta)*\sigma_j$ in the stress matrix $[P_g]$, the minimum number of stress modes in stress matrix [P] is equal to $(n_h/2=)$ 4. Using iso-function method, the initial stress matrix [P] is derived directly from the assumed displacement field. It is

$$[P]=\begin{bmatrix} 1 & 0 & 0 & \xi & 0 & 0 & \eta & 0 & 0 & \zeta & 0 & 0 & \xi\eta & 0 & 0 & \eta\zeta & 0 & \xi\zeta & 0 \\ 0 & 1 & 0 & 0 & \xi & 0 & 0 & \eta & 0 & 0 & \zeta & 0 & 0 & \xi\eta & 0 & 0 & 0 & 0 & \xi\zeta \\ 0 & 0 & 1 & 0 & 0 & \xi & 0 & 0 & \eta & 0 & 0 & \zeta & 0 & 0 & \xi\eta & 0 & \eta\zeta & 0 & 0 \end{bmatrix} \tag{4-259}$$

Then, by means of the classification method of stress modes, one obtains

$$[P]=\begin{bmatrix} 1 & 0 & 0 & 0 & 0 \\ 0 & 1 & 0 & \xi & 0 \\ 0 & 0 & 1 & 0 & \eta \end{bmatrix} \tag{4-260}$$

In the stress matrix [P], there are 5 stress modes. The result of eigenvalue examination show that there are not any kinematic deformation modes. The eigenvalues λ_i of the element are given in the Table 19.

Table 19 Eigenvalue of Stiffness Matrix for the 3-D, 8-node Hybrid Element with 10 stress modes and isotropic materials: E=1100 GPa, ν=0.3

No.	λ_i (*10^3)	No.	λ_i (*10^3)	No.	λ_i (*10^3)
1	0.09402	7	0.1813	13	0.8462
2	0.1410	8	0.2821	14	0.8462
3	0.1410	9	0.5440	15	0.8462
4	0.1410	10	0.5440	16	0.8462
5	0.1410	11	0.5641	17	0.8462
6	0.1813	12	0.7051	18	2.7500

Multilayer Element Stiffness Matrix

When N>2, the stress matrix [P] (4-260) can not be used to formulate the multilayer element because it does not contain enough stress modes. According to equation (4-258), a stress matrix [P] should contain 7 stress modes at least for a general multilayer element. In this case, the iso-function partial stress matrix (4-259) does not contain enough necessary stress modes. Therefore, more polynomial terms have to be added into the stress matrix for examining. For instance, the quadratic terms should be included. Using the classification method, the following stress matrix [P] is obtained

$$[P] = \begin{bmatrix} 1 & 0 & 0 & 0 & 0 & \xi\eta & 0 & 0 \\ 0 & 1 & 0 & \xi & 0 & 0 & \zeta^2 & 0 \\ 0 & 0 & 1 & 0 & \eta & 0 & 0 & \zeta^2 \end{bmatrix} \tag{4-261}$$

Using this stress matrix, the examination of the element indictes that the multilayer element, which consists of different sub-elements from N=1 to N=50, does not have any kinematic deformation modes.

4.3.3 Multilayer Transition Element

The 6-node partial hybrid transition elements also can be used to formulate a multilayer transition element which may connect a multilayer solid element with a degenerated plate element.

Sub-Element Matrices

The multilayer transition element consists of a stack of 6-node transition element (see figure 30). For the i-th sub-element, the assumed displacement field is the same as that for 6-node transition element in section 4.2.2. It is in the form,

$$\begin{Bmatrix} u^i \\ v^i \\ w^i \end{Bmatrix} = \sum_1^4 N_j \begin{Bmatrix} u_j^i \\ v_j^i \\ w_j^i \end{Bmatrix} + \sum_a^b N_j \left[\begin{Bmatrix} u_j^0 \\ v_j^0 \\ w_j^0 \end{Bmatrix} + \zeta' [b_j] \begin{Bmatrix} \psi_{xj} \\ \psi_{yj} \\ \psi_{zj} \end{Bmatrix} \right] \tag{4-262}$$

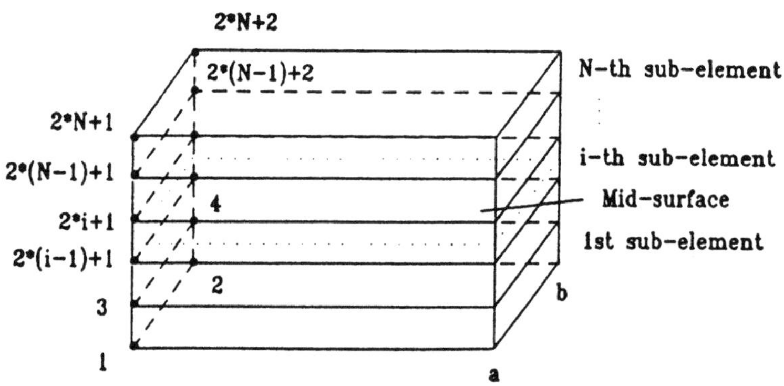

Figure 30 Multilayer transition element

The nodal dispalcement vector of the i-th sub-element can be written in the form,

$$\begin{aligned} \boldsymbol{\delta}^i &= [\boldsymbol{d}^i \ \boldsymbol{d}^{i+1} \ \boldsymbol{d}^0] \\ &= [\boldsymbol{d}_1^i \ \boldsymbol{d}_2^i \ \boldsymbol{d}_1^{i+1} \ \boldsymbol{d}_2^{i+1} \ \boldsymbol{d}_1^0 \ \boldsymbol{d}_2^0] \end{aligned} \tag{4-263}$$

where $\mathbf{d}^i$ and $\mathbf{d}^{i+1}$ are the nodal displacement vectors related to the lower and upper surfaces of the sub-element, and $\mathbf{d}^0$ is the nodel displacement vector related to the line of nodes. Within the sub-element, the partial stress field is assumed in the form,

$$\sigma_g^i=\begin{Bmatrix}\sigma_z^i\\ \sigma_{yz}^i\\ \sigma_{zx}^i\end{Bmatrix}=[P_g]\,\boldsymbol{\phi}^i=[P]\,\frac{1}{2}\{(1+\zeta)\,\alpha_T^i+(1-\zeta)\,\alpha_B^i\} \tag{4-264}$$

where the stress matrix [P] is a function consisted of even order terms of the coordinate ζ. Using the equations (4-11) and (4-147)-(4-171), the sub-element matrices can be obtained as follows,

$$\begin{aligned}
[H^i] &=-\int_{V_i}[P_g]^T[R_3^i]\,[P_g]\,dV\\
[G^i] &=\int_{V_i}[P_g]^T(\,[B_L]+[R_2^i]^T[B_g]\,)\,dV\\
[K_d^i] &=\int_{V_i}[B_g]^T[R_1^i]\,[B_g]\,dV\\
\boldsymbol{f}^i &=\int_{V_i}[N]^T\boldsymbol{F}dV+\int_{S_{ti}}[N]^T\boldsymbol{T}dS
\end{aligned} \tag{4-225}$$

Multilayer Matrices

Applying the assembling rules (4-237) and (4-244),

$$\boldsymbol{\delta}=\sum_{i=1}^{N}\boldsymbol{\delta}^i=[\boldsymbol{d}^1\ \boldsymbol{d}^2\ \ldots\ \boldsymbol{d}^{N+1}]^T \tag{4-237}$$

and

$$\boldsymbol{\varphi}=\sum_{i=1}^{N}\boldsymbol{\alpha}^i=[\boldsymbol{\alpha}^1\ \boldsymbol{\alpha}^2\ \ldots\ \boldsymbol{\alpha}^{N+1}]^T \tag{4-244}$$

the sub-element matrices from the 1st layer to N-th layer are added into the multilayer matrices,

$$[K]_d=\sum_{i=1}^{N}[K_d^i] \qquad [H]=\sum_{i=1}^{N}[H^i]$$
$$[G]=\sum_{i=1}^{N}[G^i] \qquad \boldsymbol{f}=\sum_{i=1}^{N}\boldsymbol{f}^i \tag{4-239}$$

Then, the stiffness matrix of the multilayer element can be calculated using equation (4-17),

$$[K]=[K_d]+[K_h]$$
$$[K_h]=[G]^T[H]^{-1}[G] \tag{4-17}$$

Examination of the Element

Similar to the multilayer solid element, the minimum number of stress modes in an assumed stress matrix for a multilayer transition element varies with the number of sub-elements in the multilayer elements. Using eigenvalue examination method, the rank of the displacement-formulated stiffness matrix $[K_d]$ is given for different multilayer elements with different number of sub-elements. The minimum number of stress modes in an assumed partial stress matrix is given in table 20.

Table 20 Minimum number of stress modes in the matrix $[P_g]$

N	n	r	n_d	n_h
1	24	6	10	8
2	30	6	14	10
3	36	6	18	12
.	.	.	.	.
10	78	6	46	26
.	.	.	.	.

The results of eigenvalue examination shows that the rank of semi-stiffness matrix $[K_d]$ increases by 4 when the multilayer element increases a surface. For example, there are two surfaces in a fundamental multilayer element consisted of a sub-element, and the rank of semi-stiffness matrix equals 10. There is one increased surface in the multilayer element consisted of two sub-elements, and the rank of matrix $[K_d]$ equals 14. Furthermore, there are eleven surfaces in the multilayer element consisted of ten sub-elements, and the rank of matrix $[K_d]$ is equal to 46. Thus, each increased surface in a multilayer element corresponds 4 deformation modes related to semi-stiffness matrix $[K_d]$. Meanwhile, each added surface will increase two point nodes which correspond 6 degrees of freedom in the multilayer element. Thus, if a multilayer element contains N layers, it will have N+1 surfaces and one has

$$n = 12+6 * (N+1) \qquad n_d = 2+4 * (N+1) \qquad r = 6 \tag{4-264}$$

Define that m' is the number of stress modes in matrix [P] related to a surface. Thus, the total number of stress modes is

$$m = (N+1)\, m' \tag{4-256}$$

The necessary and sufficient condition (4-44) for avoiding kinematic deformation modes is

$$(N+1)\, m' = 12+6*(N+1) -2 -4*(N+1) - 6 \tag{4-265}$$

Thus, one obtains minimum number of stress modes in the matrix [P] related to each surface for the multilayer element as follows,

$$m' = 2+\frac{4}{N+1} \tag{4-266}$$

Using this formulation, one can calculate the number of stress modes in the stress matrix [P] related to each surface for multilayer element consisted of different number of sub-elements.

N = 1	m' = 4
N = 2	m' = 4
N = 3	m' = 3
N = 4	m' = 3
N = 5	m' = 3
N = 6	m' = 3
....	....

$$N = 100 \qquad\qquad m' = 3$$

Therefore, the number of stress modes needed in a sub-element in order to avoid kinematic deformation modes is different for different N.

Sub-Element Stiffness Matrix and Kinematic Deformation Modes

When N=1, the multilayer element becomes a sub-element. The sub-element is a 6-node partial hybrid transition element. The examination of the element has been given in section 4.2.4. For completeness, the analysis is given here again. The sub-element has (n=) 24 degrees of freedom and (r=) 6 degrees of rigid motion. Thus, the sub-element has 18 natural deformation modes. The eigenvalue examination indicates that the partial stiffness matrix $[K_d]$ gives 10 non-zero eigenvalues and represents 10 natural deformation modes of the element. Therefore, another partial stiffness matrix $[K_h]$ must give 8 non-zero eigenvalues and represent 8 natural deformation modes. Thus, the minimum number of the stress modes in the assumed stress field $[P_g]$ is equal to 8. Due to the fact that a stress mode σ_j in the stress matrix [P] represents two stress modes in the stress matrix $[P_g]$, the minumum number of stress modes in stress matrix [P] is equal to ($n_h/2$=) 4. Using iso-function method, the initial stress matrix [P] is derived directly from the assumed displacement field. It is

$$[P] = \begin{bmatrix} 1 & 0 & 0 & \xi & 0 & 0 & \eta & 0 & 0 & \zeta & 0 & 0 & \xi\eta & 0 & 0 & \eta\zeta & 0 & \xi\zeta & 0 \\ 0 & 1 & 0 & 0 & \xi & 0 & 0 & \eta & 0 & 0 & \zeta & 0 & 0 & \xi\eta & 0 & 0 & 0 & 0 & \xi\zeta \\ 0 & 0 & 1 & 0 & 0 & \xi & 0 & 0 & \eta & 0 & 0 & \zeta & 0 & 0 & \xi\eta & 0 & \eta\zeta & 0 & 0 \end{bmatrix} \tag{4-267}$$

Then, by means of the classification method of stress modes, one obtains an optimal stress matrix,

$$[P] = \begin{bmatrix} 1 & 0 & 0 & 0 & 0 \\ 0 & 1 & 0 & \xi & 0 \\ 0 & 0 & 1 & 0 & \eta \end{bmatrix} \tag{4-268}$$

In the stress matrix [P], there are 5 stress modes. The result of eigenvalue examination show that there are not any kinematic deformation modes.

Multilayer Element Stiffness Matrix and Locking Phenomenon

When N=2, the stress matrix [P] (4-268) can be used to formulate the multilayer element. But when N≥3, the stress matrices [P] can not be used to formulate a multilayer transition element due to the fact that the locking phenomenon appears.

The locking means that the solution becomes zero when the plate element becomes thin. This phenomenon appears in C^0 finite element because the Kirchhoff constraint can not be satisfied when plate element becomes thin. In the multilayer transition element, the sub-element will become thin when their number within a fixed thickness multilayer element increases. For the multilayer transition element, equal order interpolation is used for lines of nodes which are used to meet with plate elements. Therefore, when sub-element becomes thinner and thinner, two spurious constraints produce the locking action on Ψ_{xi} and Ψ_{yi} . In order to remove the locking phenomenon, several methods can be used such as unequal order interpolation, reduced integration, assumed strain approach, additional incompatible modes, field-redistribution, and so on [4.44]. In this work, the advantage of hybrid stress finite element is used to overcome locking phenomenon in the element as follows:

By calculating m' (4-266), it has been shown that the minimum number of stress modes in the stress matrix [P] decreases to 3. Therefore, there are unnecessary stress modes in the matrix [P] (4-268) for multilayer transition elements ($N \geq 3$) and the extra stress modes in the stress matrix [P] results in over-stiffness and lead to locking phenomenon. The classification method gives following stress matrix that can be used to avoid locking phenomenon are

$$[P] = \begin{bmatrix} 1 & 0 & 0 \\ 0 & 1 & 0 \\ 0 & 0 & 1 \end{bmatrix} \tag{4-269}$$

The results of the eigenvalue examination for the multilayer element with 3 sub-elements are given in the table 21.

Other multilayer elements with different number of sub-elements also can be examined. The examination of the multilayer elements shows that there is no spurious constraints in the multilayer element when the stress matrix [P] (4-269) is used.

Table 21 Eigenvalue of Stiffness Matrix for Hybrid Multilayer Element with 3 sub-elements and isotropic materials: E=1100 GPa, ν=0.3

No.	λ_i (*10^3)	No.	λ_i (*10^3)	No.	λ_i (*10^3)
1	0.00060	11	0.1602	21	0.9849
2	0.00141	12	0.2971	22	1.0190
3	0.00186	13	0.3059	23	1.1640
4	0.00239	14	0.4109	24	1.2140
5	0.00967	15	0.5011	25	1.3130
6	0.03671	16	0.5534	26	1.5580
7	0.04086	17	0.6002	27	1.6150
8	0.08544	18	0.6734	28	2.0050
9	0.1077	19	0.7466	29	2.9820
10	0.1393	20	0.8111	30	4.2070

REFERENCES

4.1 Hoa S.V. and W. Feng, 'Finite Elements for Analysis of Composite Structures', in Ed., S.V. Hoa, Computer-Aided Design of Polymer-Matrix Composite Structures, Marcel Dekker, Inc., 1995.

4.2 A.K. Noor, 'Mechanics of anisotropic plates and shells-a new look at an old subject', Computers & Structures, vol. 44, no.3, 499-514(1992).

4.3 J.N. Reddy & D.H. Robbins Jr., 'Theories and computational models for composite laminates', Appl. Mech. Rev., vol. 47, no. 6, part 1, 147-169(1994).

4.4 E.F. Rybicki, 'Approximate three-dimensional solution for symmetric laminates under in-plane loading', J. Compo. Mater., vol. 5, p354(1971).

4.5 L.B. Lessard, M.M. Shokrieh & A.S. Schmidt, '3-D stress analysis of composite plates with or without stress concentrations', Composites Modelling and Processing Science, III, ICCM/9, Ed. Antonio Miravete, Woodhead Publishing Limited, (1993).

4.6 R.M. Barker, F.T. Lin and J.R. Dana, '3-D finite element analysis of laminated composites', Computers & Structures, vol.2, 1013-1029(1972).

4.7 A.K. Noor and W.S. Burton, 'Assessment of shear deformation theories for multilayered composite plates', Appl. Mech. Rev., vol. 42, no. 1, 1-9(1989).

4.8 E. Reissner and Y. Stavsky, 'Bending and stretching of certain types of heterogeneous aeroisotropic elastic plates', J. Appl. Mech. ASME, 28, 402-408(1961).

4.9 K. H. Lo, R.M. Christensen and E.M.Wu, 'A higher-order theory of plate deformation: Part 1, Homogeneous plates; Part 2, Laminated plates', J. Appl. Mech. ASME, 44, 663-676(1977).

4.10 J. N. Reddy, 'A simple higher-order theory for laminated composite plates', J. Appl. Mech. ASME, 51,745-752(1984).
4.11 J.N. Reddy, 'A refined nonlinear theory of plates with transverse shear deformation', Int. J. Solids Struct., 20, 881-896(1984).
4.12 J. M. Whitney and C.T. Sun, 'A higher order theory for extensional motion of laminated composites', J. Sound Vib., vol. 30, 85-97(1973).
4.13 M. Epstein and H.P.Huttelmaier, 'A finite element formulation for multilayered and thick plates', Comp. Struct., 16, 645-650(1983).
4.14 R. L. Hinrichsen and A.N. Palazotto, 'Nonlinear finite element analysis of thick composite plates using a cubic spline function', AIAA J., 24, 1836-1842(1986).
4.15 A. Muc, 'Computational models and variational formulations for laminated composite structures', Composites Modelling and Processing Science, ICCM/9, vol. III, ed., Antonio Miravete, University of Zaragoza, Woodhead Publishing Limited (1993).
4.16 D. H. Robbins, Jr. and J.N. Reddy, 'Modelling of thick composites using a layer-wise laminate theory', Int. J. Numer. Methods Eng., vol.36, 655-677(1993).
4.17 S. V. Hoa and W. Feng, 'Finite element method for composites', Composites '96 and Oriented Polymers Symposium, Montreal, Canada (1996).
4.18 S. V. Hoa and W. Feng, 'Development of hybrid finite elements for stress analysis of composite structures', the 3rd International Conf. on Fracture and Strength of Solids, Hong Kong, China (1997).
4.19 Q. Huang, 'Three Dimensional Composite Finite Element for Stress Analysis of Anisotropic Laminate Structures', Ph. D. Dissertation, Concordia University, Montreal, Canada (1989).
4.20 J. Han, 'Three dimensional multilayer composite finite element for stress analysis of composite laminates', Ph.D. Dissertation, Concordia University, Montreal, Canada, (1994).
4.21 W. Feng and S.V. Hoa, 'A 3-D partial hybrid laminated element for analysis of thick laminates', Third Int. Conf. on Composites Engineering, New Orleans, USA (1996).
4.22 S. Ahmad, B. M. Irons and O.C. Zienkiewicz, 'Analysis of thick and thin shell structures by curved finite elements', Int. J. Numer. Methods Eng., vo. 2, 419-451(1970).
4.23 W. C. Chao and J. N. Reddy, 'Analysis of laminated composite shell using a degenerated 3-D element', Int. J. Numer. Methods Eng., vo. 20, 1991-2007(1984).
4.24 J. Blocki, 'A high-order linear theory for isotropic plates - I. theoretical considerations', Int. J. Solid Structures, vol. 29, no. 7 825-836(1992).
4.25 W. Feng and S. V. Hoa, 'A degenerated plate/shell element with assumed partial stress field for the analysis of laminated composites', 4th International Conf. on Computer Aided Design in Composite Material Technology, Southampton, UK (1994).
4.26 W. Feng and S.V. Hoa, 'A partial hybrid degenerated plate/shell element for the analysis of laminated composites', Int. J. Numer. Methods Eng., vol. 39, 3625-3639(1996).
4.27 A. K. Noor, "Global-local methodologies and their application to nonlinear analysis", Finite Elements in Analysis and Design, vol. 2, 333-346(1986).
4.28 K. J. Bathe, Finite Element Procedures in Engineering Analysis, Prentice-Hall, N.J., 1982.
4.29 K. S. Surana, "Geometrically nonlinear formulation for the three dimensional solid-shell transition finite elements", Comp. Struct., vol. 15, 549-566(1982).'
4.30 C. L. Liao, J.N. Reddy and S.P. Engelstad' "A solid-shell transition element for geometrically non-linear analysis of laminated composite structures", Int. J. Numer. Methods in Eng., vol. 26, 1843-1854(1988).
4.31 W. Feng and S.V. Hoa, '3-D transition element formulation for the global-local analysis of laminated structures', Int. Conf. on Design and Manufacturing Using Composites, Montreal, Canada (1994).
4.32 S.V. Hoa and W. Feng, 'A global/local model for analysis of composites', 10th International Conf. on Composite Materials, Whistler, Canada (1995).
4.33 S. V. Hoa and W. Feng, 'Global/local approach using hybrid elements for composites', 5th International Conf. on Computer Aided Design in Composite Material Technology, Udine, Italy (1996).
4.34 S. V. Hoa and W. Feng, 'A global/local model using partial hybrid elements', XIXth International Conf. of Theoretical and Applied Mechanics, Kyoto, Japan (1996).
4.35 S.V. Hoa and W. Feng, 'Application of a global/local finite element model to composite laminates', Science and Engineering of Composite Materials, vol. 5, 151-168(1996).

4.36 W. Feng and S. V. Hoa, 'Partial hybrid finite elements for composite laminates', Finite Elements in Analysis and Design, (in press).
4.37 K. J. Han and P.L. Gould, "Line node and transitional shell element for rotational shells", Int. J. Numer. Methods in Eng., vol. 18, 879-895(1982).
4.38 O. C. Zienkiewicz, The Finite Element Method, 3rd eds, McGraw-Hill, New York, 1977.
4.39 H. Kardestuncer and D. H. Norrie, Finite Element Handbook, McGraw-Hill, 1987.
4.40 J.-H. Han and S.V. Hoa, 'A three-dimensional multilayer composite finite element for stress analysis of composite laminates', Int. J. Numer. Methods Engrg., vol.36, 3903- 3914(1993).
4.41 W. Feng and S.V. Hoa, 'A multilayer element with partial assumed stress field for analysis of laminated structures', The 16th Canadian Congress of Applied Mechanics CANCAM 97, Quebec, Canada (1997).
4.42 R.L. Spilker, ' A hybrid stress finite element formulation for thick multilayer laminates', Computers & Structures, vol. 11, 507-514 (1980).
4.43 R. L. Spilker, 'Hybrid-stress eight-node elements for thin and thick multilayer laminated plates', Int. J. Numer. Methods Eng., vol. 18, 801-828 (1982).
4.44 G. Prathap, The Finite Element Method in Structural Machanics, Kluwer Academic Publishers, Netherlands, 1993.

Chapter 5

NUMERICAL EXAMPLES OF FINITE ELEMENT ANALYSIS AND GLOBAL/LOCAL APPROACH

5.1 INTRODUCTION

Finite element method has been widely used in the analysis of structures because of the power of the technique and also because of the availability of many commercial finite element programs. Finite element analysis is a numerical analysis of the mathematical models used to represent the behaviour of engineering structures. Therefore, mathematical assumptions concerning the representation of the geometry and behaviour of the structures have to be made in finite element models. In order to efficiently and accurately perform the finite element analysis of a composite structure, it is necessary to have a qualitative knowledge of the structure behaviour and its finite element model.

To perform a finite element analysis of a structure, as mentioned in chapter 1, the structure must be discretized into a set of elements, which are quasi-disjoint non-overlapping elements. These elements are connected by a set of nodes. The collection of nodes and elements forms a finite element mesh. A variety of element types are available today. The analyst or designer can mix element types to solve one problem. It should be noted that the choice of element types and element mesh is problem-dependent. The number of nodes and the type of elements to be used in a finite element model is a matter of engineering judgment. As a general rule, the larger is the number of nodes and elements, the more accurate is the finite element solution, but also the more expensive the solution is. More memory space is needed to store the finite element model, and more computer time is needed to obtain the solution.

In practice, most composite laminates contain local regions where thick conditions prevail throughout. For example, the presence of an open hole in a laminate introduces significant transverse stresses which create a very complicated

3-D stress field in the vicinity of the hole. This complex state of stress depends on the stacking sequence of the laminate, the fibre orientation of each lamina, and the material properties of the fibre and of the matrix. In order to obtain the stress fields in these localities, a detail 3-D finite element analysis is required. However, a detailed full 3-D analysis of these laminates to obtain accurate stresses may require a huge number of nodes and elements. They may exhaust the computer resources.

In order to keep the number of nodes and elements down, one way is to classify the domain of the structure to be analyzed into different regions. In different regions, mesh densities and element types vary [5.1-5.3]. In general, the region where large gradients of displacements and/or stresses to be expected is discretized into many elements (fine mesh); otherwise, the region is modeled using few elements (coarse mesh). For example, if a structure contains a crack, the local region near the crack is usually divided into many "very small" elements in order to predict accurately stress distribution and other region is discretized into "large" elements to model the response of the structure (see figure 31).

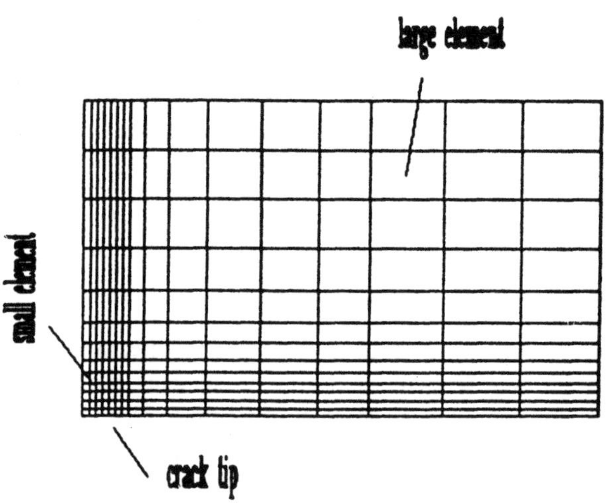

Figure 31 An element mesh with different mesh densities

Generation of element meshes with single element type is relatively easy due to the compatibility of elements with the same degree of freedom. The element mesh for the stress analysis of a structure can be refined using two approaches: h-version and p-version refinement scheme [5.4-5.6]. The h-version scheme is to subdivide a selected element into a number of smaller elements of the same type. The discretization is improved by reducing the element size. The p-version scheme is to replace a selected element by an element of higher order. The discretization is improved by increasing the polynomial order of an existing element.

For stress analysis of composite laminates, analysts and designers have to use different element types to solve a practical problem due to the complex nature of composite laminates. In finite element method, the stress analysis of composite laminates is done using three techniques: 3-D solid modelling, layerwise modelling and equivalent single layer with smeared properties. In 3-D solid modelling [5.7], no specific kinematic assumptions are introduced regarding the behaviour of the structure. It takes the behaviour of the individual laminae into consideration. However, since the composite laminae are usually very thin, the 3-D element usually runs into problem due to the large aspect ratios of the elements. In addition, refining the finite element mesh can quickly exhaust the computer storage. In the layer-wise modelling [5.8], the individual laminae are taken as 2-D layers. These layers are then assembled through the thickness. Its advantage is that it requires only 2-D finite element mesh, and the element aspect ratio is restricted to 2-D considerations. In practice, a typical composite may have many layers, each of which requires one 2-D layer through the thickness. The number of degrees of freedom per node is directly proportional to the number of layers in a laminate. This increases drastically the number of unknowns in a finite element model. Hence, this type of modelling is also computationally expensive. In the equivalent single layer models [5.9], the variations in orientation and properties across the thickness are integrated to obtain single properties across the thickness. This element can be used for problems such as vibration or buckling but do not provide useful results if interlaminar stresses are required. Therefore, the finite element model using same element type is not efficient for stress analysis of composite laminates. It is necessary to combine elements of different types in one finite element model to solve a problem. This kind of finite element model is referred to as a global/local finite element model.

A wide variety of global/local models have been proposed [5.1, 5.10-5.11]. In general, the global/local finite element models can be classified into two classes: sequential and simultaneous global/local models. The main difficulty with these models is the maintenance of displacement continuity along boundaries separating regions. In the sequential global/local model, the domain of the structure to be analyzed is classified into two regions: local and global. The first step in performing a global/local analysis is to set up an economical, yet adequate model by using a coarse mesh to determine the structural response. Then, the resulting displacements or stresses are imposed on the boundary of the local region for a subsequent local analysis by using a finer mesh. In order to take account of the effect of the local region on the global region, the iterative methods have to be used to establish equilibrium or compatibility along the global/local boundary. It requires much computing time. In the simultaneous global/local model, the domain of the structure to be analyzed is classified into three regions: local, global, and transition as shown in Figure 32. Each region requires an appropriate type of element for modelling the structure. This model does not require reanalysis and saves computer time. But in transition region it is necessary to use a special transition element which has different number of nodes on different sides of the element. For example, a special element is needed (see figure 33) to connect a 2-D linear element with a 2-D quadratic element.

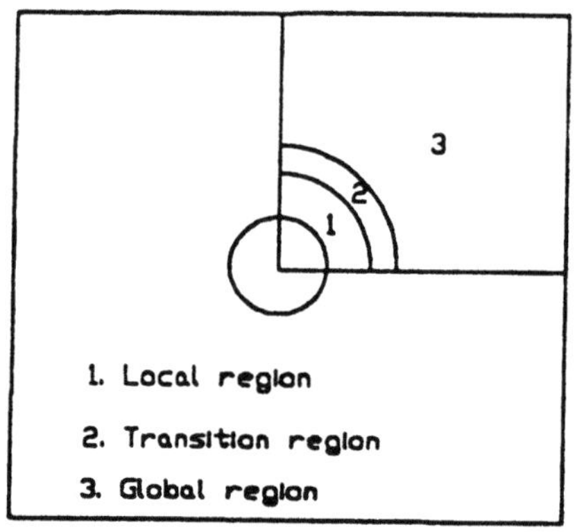

Figure 32 Global/local finite element model

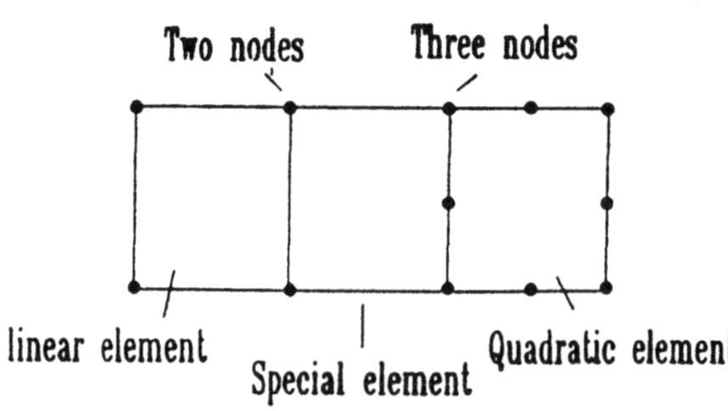

Figure 33 A special element between linear and quadratic elements

In this chapter, the partial hybrid elements presented in chapter 4 are used to establish finite element models of composite laminates and develop simultaneous global/local finite element approaches for stress analysis of composite laminates.

5.2 FINITE ELEMENT ANALYSIS USING DEGENERATED PLATE ELEMENT [5.12-5.13]

The accuracy of the finite element model using partial hybrid degenerated plate elements is demonstrated by studying the behaviour of a square laminated plate and a long laminated strip. The two laminates are ideal structures to verify the degenerated element since closed-form elasticity solutions are available. Once the accuracy of the element is verified, the element can be used to develop global/local models for stress analysis of composite laminates.

Example 1. Deflection of a Square Laminate Subjected to Uniform Loading

A three-ply square laminate with identical top and bottom plies is analyzed by using the 8-node degenerated element. Each layer in the laminate is idealized as a homogeneous orthotropic material. The relative values of the moduli in the principal material coordinate system are the same in all the plies as follows,

$$
\begin{aligned}
&E_2 / E_1 = 0.525000 && E_3 / E_1 = 0.569399 \\
&G_{12} / E_1 = 0.292813 && G_{23} / E_1 = 0.297133 \qquad (5-1) \\
&G_{13} / E_1 = 0.178088 && \upsilon_{12} = 0.440462 \\
&\upsilon_{23} = 0.180666 && \upsilon_{13} = -0.061321
\end{aligned}
$$

An uniform loading q_0 acts on the top of the simply supported laminate. The dimensions of the plate are a, b (=a) and thickness H (= 0.1a). The thickness of the top and bottom plies h_1 is equal to 0.1H, and the thickness of the middle ply h_2 is equal to 0.8H. By means of the symmetry of the problem, only one quadrant of the plate is modeled ($0<x<a/2$, $0<y<b/2$, $0<z<H$). The computational domain is modeled using 2×2 uniform meshes (see figure 34). For this particular problem, a 3-D exact solution was presented by Srinivas and Rao [5.14]. The results of the deflection w at the centre of the laminate are given in the Table 22. In the Table, E_{1t} is the modulus of the top and bottom plies and E_{1m} is the modulus of the middle ply. E_{x2} is a parameter which can be calculated from the material constants of the middle ply. In the calculation, E_1 is equal to 0.8979495×10^6, and then E_{x2} is equal to 10^6. The present solutions of the centre deflection are close to the 3-D elasticity solutions.

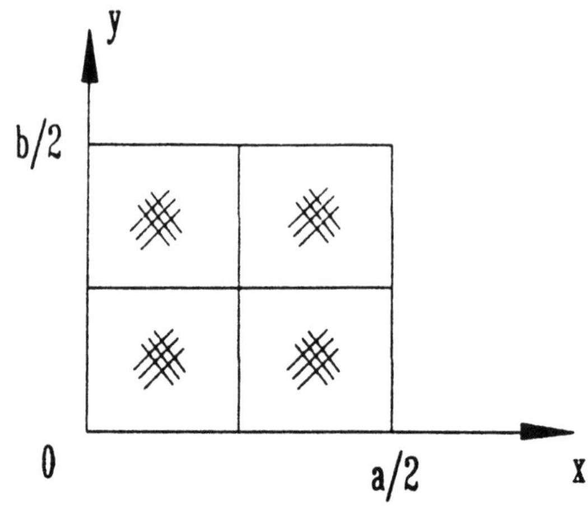

Figure 34 2x2 uniform mesh of a quadrant of the laminate

Table 22 Deflection of the simply supported laminate subjected to uniform loading

E_{1t} / E_{1m}		1	5	10	15	20
$\frac{wE_{x2}}{Hq_0}$	Exact Solution	688.58	258.97	159.38	121.72	---
	Present Solution	693.91	261.36	162.27	123.78	102.98

Example 2. Bending of a Square Laminate

The degenerated element is also used to analyze a square, simply supported, laminated plate with the [0,90,0] layers of equal thickness. Each layer of the laminated plate is also idealized as a homogeneous orthotropic material with the following material coefficients in the principal material coordinate system:

$$E_L=172.4 \text{ GPa} \qquad E_T=6.90 \text{ GPa} \qquad \upsilon_{LT}=\upsilon_{TT}=0.25$$

$$G_{LT}=3.45 \text{ GPa} \qquad G_{TT}=1.38 \text{ GPa} \qquad (5-2)$$

The upward transverse load is distributed on the top surface,

$$q(x,y) = q_0 \sin\left(\frac{\pi x}{a}\right) \sin\left(\frac{\pi y}{b}\right) \tag{5-3}$$

The dimensions of the plate are a, b and thickness H. The ratio S is defined as a/H. Due to the symmetry of the problem, only one quadrant of the plate is modeled (0<x<a/2, 0<y<b/2, 0<z<H). The computational domain is modeled using 2x2 uniform meshes. For this particular problem, the solution exists using 3-D elasticity theory[5.15] and classical laminate theory (denoted CLT). The CLT solution for τ_{xz} was found by the equations of equilibrium as discussed in [5.16]. The solutions for thick plate S=4 are given in figures 35-36. Each function is plotted along the vertical line on which it assumes its maximum value. The following normalized quantities are defined,

$$(\overline{\tau_{xy}}) = \frac{1}{q_0 S^2}(\tau_{xy}) \qquad \overline{\tau_{xz}} = \frac{1}{q_0 S}\tau_{xz} \qquad \overline{W} = \frac{100 E_T W}{q_0 H S^4} \tag{5-3}$$

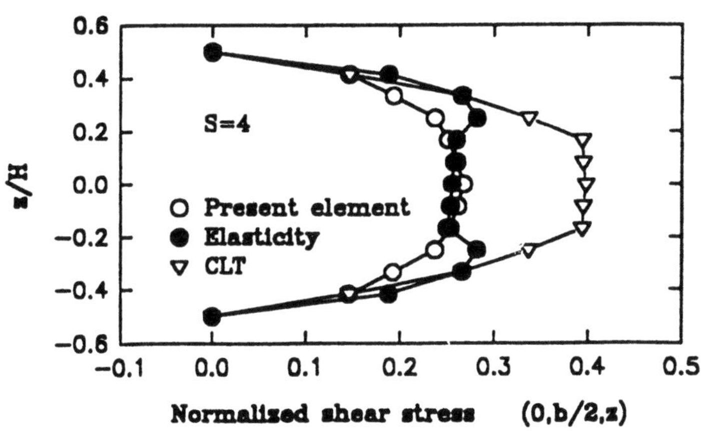

Figure 35 Normalized transverse shear stress $\overline{\tau}_{xz}$ distribution (a=b)

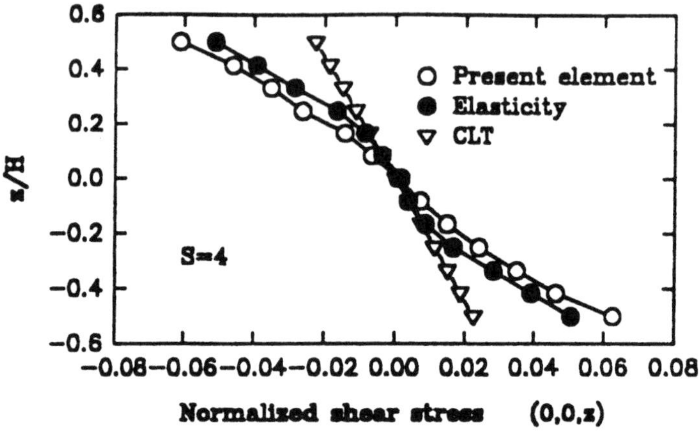

Figure 36 Normalized in-plane shear stress $\bar{\tau}_{xy}$ distribution (a=b)

The performance of finite element analysis only takes 2.03 seconds CPU time on VAX 6510 Computer by using the degenerated element. The degenerated finite element solutions are close to the exact three-dimensional elasticity solutions shown in figures 35-36 for the shear stresses.

Example 3. Cylindrical Bending of A Laminated Strip

Two infinitely long laminated strips with layers of equal thickness are simply supported along the two edges and is subjected to sinusoidal transverse load of intensity q_0

$$q(x) = q_0 \sin\left(\frac{\pi x}{l}\right) \tag{5-5}$$

The lamina material properties are the same as in example 2. Because the laminate is quite long in y direction, the displacement gradients can be neglected with respect to the y coordinate. Hence, a slice which is taken out from the structure was modeled. Because of symmetry, numerical analysis is carried out over one half of the slice and it is subdivided into 2 equal elements. This problem has an elasticity solution[5.16] and a CLT solution. Pian and Li[5.17] also calculated stresses for this problem using a 14 DOF, 2-D partial hybrid element. For 3-layer laminate [0,90,0], the maximum central deflection as a function of span-to-depth ratio is plotted in figure 37. The result is in agreement with the elasticity solution. It takes 1.55 seconds CPU time to solve the problem on the VAX 6510 computer. For the 20-layer laminate $[90,0]_{10T}$, the result of the transverse shear stress which is normalized by the applied load q_0 is also in agreement with the elasticity solution as shown in fig. 38.

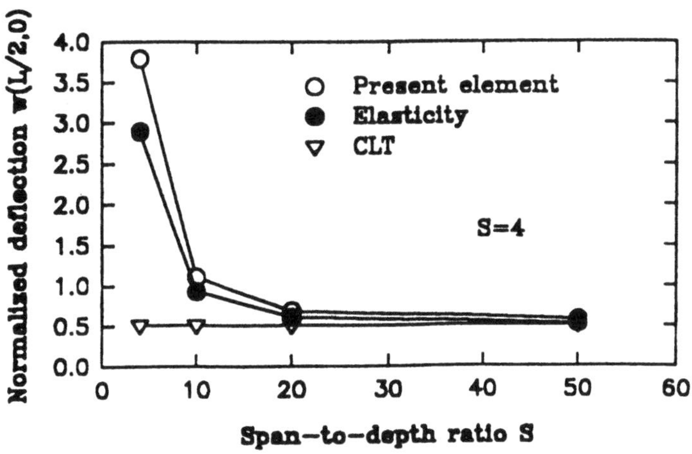

Figure 37 Maximum central deflection as function of span-to-depth ratio

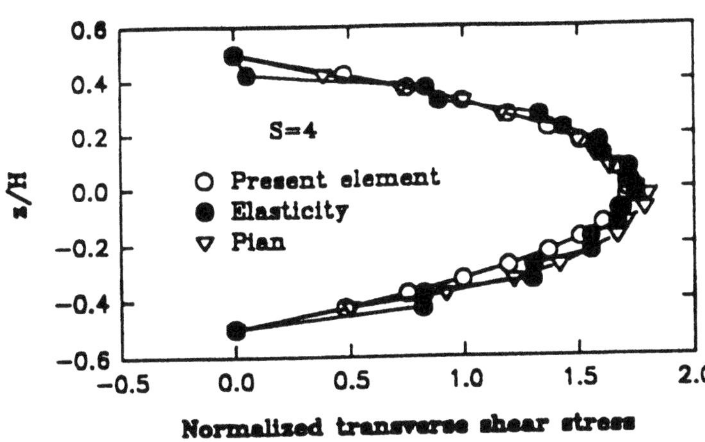

Figure 38 Shear stress distribution at edge of 20-layer $[90,0]_{10T}$ laminate

In the examples, it has been shown that finite models using the degenerated element is computationally most efficient for the stress analysis of composite laminates. It can provide accurate solutions for the deflection of laminates. It also can predict stresses accurately in the laminates with large number of layers. However it can not be used to evaluate stress concentration because of the limitations of assumed displacement field over the whole thickness. Therefore, the element can adequately describe the global region in the global/local stress approach which will be presented in this chapter.

5.3 FINITE ELEMENT ANALYSIS USING SOLID ELEMENT [5.18-5.20]

In order to predict stress distribution accurately in the local region where high stress gradient exists, 3-D solid element or multilayer element is needed. In this section, a long laminated strip subjected to bending loads is investigated using 3-D, 8-node element to verify the efficiency and accuracy of the partial hybrid solid element.

The three-layer laminated strip with fibre orientation [0,90,0] is supposed to be infinitely long in the y direction and simply supported along the two edges x=0 and l (see figure 39). On the top surface, it is subjected to sinusoidal transverse load of intensity q_0. The loading function is given in equation (5-5).

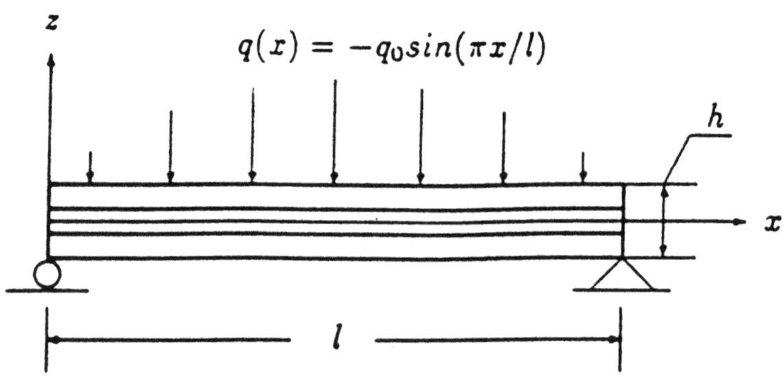

Figure 39 The cross section of infinitely long laminated strip [0,90,0] subjected to distributed transverse loading

The lamina material properties in the principle material direction are

$$E_L=171 \text{ GPa} \qquad E_T=3.42 \text{ GPa} \qquad \upsilon_{LT}=\upsilon_{TT}=0.25$$

$$G_{LT}=3.42 \text{ GPa} \qquad G_{TT}=1.37 \text{ GPa} \tag{5 - 6}$$

Similar to the example 3 in section 5.2, a slice was taken out from the laminated strip for establishing the finite element model. Because of symmetry, finite element analysis is carried out over half of the slice. There are ten uniform elements in the half length along the x-direction, one element in the y direction, and two, four, eight elements in each layer for three finite element meshes which has a total of 60, 120 and 240 elements, respectively. The numerical results are presented in terms of normalized values which are defined as

$$\overline{\sigma_x}=\frac{\sigma_x}{q_0} \qquad \overline{\tau_{xz}}=\frac{\tau_{xz}}{q_0}$$

$$\overline{u}=\frac{100E_T h^3 u}{q_0 L^4} \tag{5-7}$$

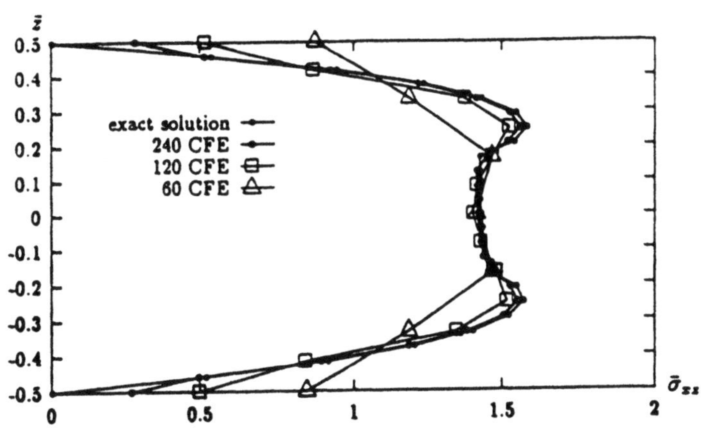

Figure 40 Stress $\overline{\tau}_{xz}$ (x=0) obtained from partial hybrid element model

The distribution of transverse stress $\overline{\tau}_{xz}$ is shown in figure 40, which compares the results calculated with 60, 120, and 240 partial hybrid elements and the results of Pagano's elasticity solution [5.16]. The comparison of stress $\overline{\sigma}_z$ and $\overline{\sigma}_x$ are also shown in figure 41 and figure 42, respectively. The distribution of displacement u is shown in figure 43.

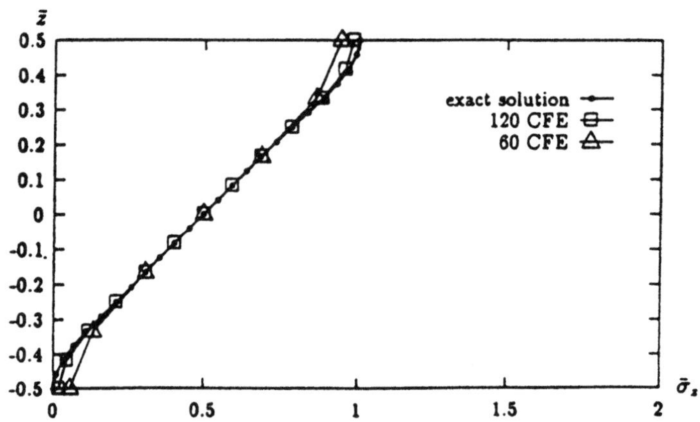

Figure 41 Stress $\overline{\sigma}_z$ (x=l/2) obtained from partial hybrid element model

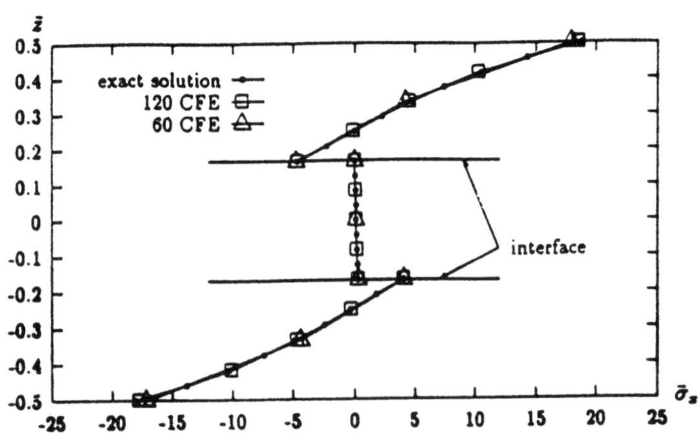

Figure 42 Stress $\overline{\sigma}_x$ (x=l/2) obtained from partial hybrid element model

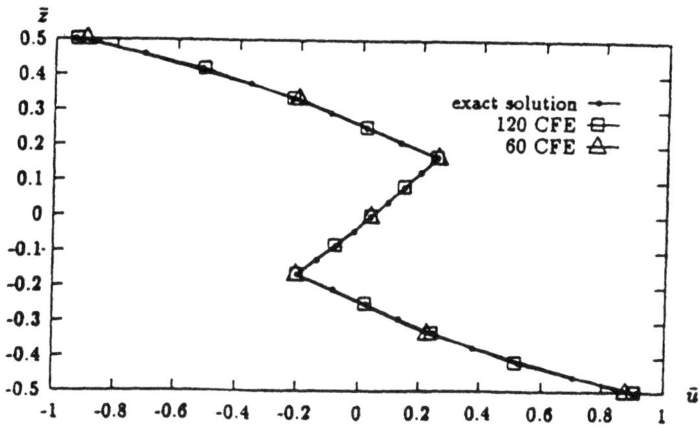

Figure 43 Displacement $\overline{u}$ (x=0) obtained from partial hybrid element model

This problem has been solved by using 432 3-D, 20-node displacement elements [5.21]. The results of the shear stress $\overline{\tau}_{xz}$ obtained from the 240 8-node partial hybrid elements and 432 20-node displacement elements are given in figure 44, compared with the result of Pagano's elasticity solution.

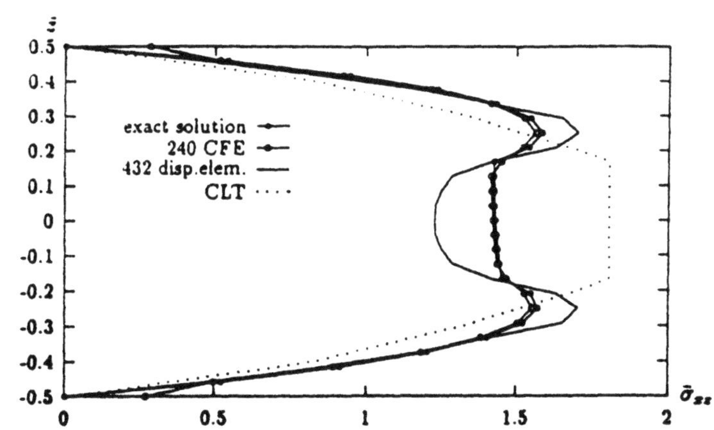

Figure 44 Stress $\overline{\tau}_{xz}$ (x=0) obtained from different finite element models

The results in figure 40 indicates that the shear stress τ_{xz} calculated by the partial hybrid finite element method quickly converges to the exact elasticity solution as the number of elements increases. Figure 44 shows the partial hybrid element solution is in better agreement with the exact Pagano's elasticity solution, although it used only 240 8-node elements while the displacement element solution uses 432 20-node elements. The CPU time consumed for 240 8-node partial hybrid elements is 2 minutes and 58.94 seconds and that for 432 20-node displacement elements is 18 minutes and 2.87 seconds on the VAX 6510 computer. Therefore, partial hybrid solid element is accurate and more efficient for stress analysis of composite laminates.

5.4 FINITE ELEMENT ANALYSIS USING MULTILAYER ELEMENT [5.19, 5.22-.23]

The accuracy of the partial hybrid multilayer element is demonstrated by calculating the stress state in a rectangular laminated plate subjected to a distributed loading and a laminated strip subjected to a bending load. These cases are selected because there are elasticity exact solutions of these problems for comparison with finite element solutions.

Example 1. Bending of a Rectangular Laminated Plate

The problem to be solved is a rectangular, simply supported, laminated plate with the [0,90,0] layers of equal thickness (see figure 45). Each layer of the laminated plate is also idealized as a homogeneous orthotropic material with the following material coefficients in the principal material coordinate system:

$$E_L=174.6 \text{ GPa} \qquad E_T= 7 \text{ GPa} \qquad \upsilon_{LT}=\upsilon_{TT}=0.25$$

$$G_{LT}= 3.5 \text{ GPa} \qquad G_{TT}= 1.4 \text{ GPa} \qquad (5 - 8)$$

where L refer to the direction parallel to the fibres and T is the transverse direction. The upward transverse load is distributed on the top surface. The loading function is given by equation (5-3).

The dimensions of the laminate are a, b and thickness H. The ratio S is defined as a/H. Due to the symmetry of the problem, only one quadrant of the plate is modeled ($0<x<a/2$, $0<y<b/2$, $0<z<H$). The computational domain is modeled with 4×4 uniform meshes. Each multilayer element consists of twelve 3-D, 8-node sub-elements through the thickness of the laminate. This particular problem has been investigated by Pagano using 3-D anisotropic elasticity theory[5.15], by Reddy using higher order shear deformation theory [5.24], and by Liou and Sun using hybrid finite element method [5.25]. The results of the analysis are presented in Tables 23

and 24. The normalized quantities are defined as

$$(\overline{\sigma_x}, \overline{\sigma_y}, \overline{\tau_{xy}}) = (\sigma_x, \sigma_y, \tau_{xy}) / (q_0 S^2)$$

$$(\overline{\tau_{yz}}, \overline{\tau_{xz}}) = (\tau_{yz}, \tau_{xz}) / (q_0 S) \tag{5-9}$$

$$\overline{w} = \frac{100 E_T w}{q_0 H S^4} \qquad \overline{z} = z/H \qquad S = a/H$$

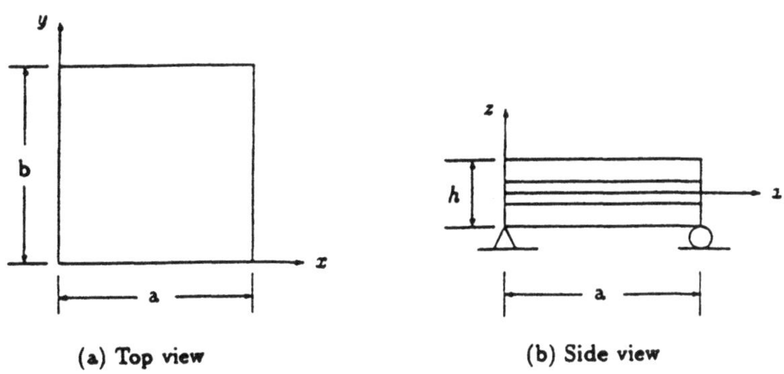

Figure 45 Bending of a simply supported rectangular laminated plate

In the tables 23 and 24, the results show that the present multilayer element provides stresses and deflection accurately. The CLT solution is accurate only for thin plate. When the span-to-thickness ratio S becomes small, the disagreement is significant.

Table 23 Normalized deflection and stresses in a square laminate (b=a)

S	Source	$\bar{\sigma}_x$(a/2, b/2, ±h/2)	$\bar{\sigma}_y$(a/2, b/2, ±h/2)	$\bar{\tau}_{xz}$(0, b/2, 0)	$\bar{\tau}_{yz}$(a/2, 0, 0)	$\bar{w}$(a/2, b/2, 0)
4	Pagano	0.801	0.534	0.256	0.217	-
		-0.755	-0.556			
	Hybrid	0.717	0.517	0.263	0.221	2.020
		-0.679	-0.541			
	Reddy	0.7346			0.1832	1.9218
	Present	0.806	0.538	0.262	0.220	2.044
		-0.760	-0.561			
10	Pagano	0.590	0.285	0.357	0.1228	-
		-0.590	-0.288			
	Hybrid	0.580	0.285	0.367	0.127	0.7548
		-0.580	-0.289			
	Reddy	0.5684			0.1033	0.7125
	Present	0.590	0.283	0.360	0.126	0.7592
		-0.589	-0.287			
20	Pagano	±0.552	±0.210	0.385	0.0938	-
	Hybrid	±0.553	±0.210	0.395	0.0998	0.5170
	Present	±0.552	±0.210	0.385	0.0971	0.5167
	CLT	±0.539	±0.180	0.395	0.0823	

Pagano: elasticity exact solution [5.15].
Hybrid: hybrid finite element method [5.25].
Reddy: high order shear deformation theory [5.24].
Present: partial hybrid multilayer element method.
CLT classical lamination theory

Table 24 Normalized deflection and stresses in a rectangular laminate (b=3a)

S	Source	$\bar{\sigma}_x$(a/2, b/2, ±h/2)	$\bar{\sigma}_y$(a/2, b/2, ±h/2)	$\bar{\tau}_{xz}$(0, b/2, 0)	$\bar{\tau}_{yz}$(a/2, 0, 0)	$\bar{w}$(a/2, b/2, 0)
4	Pagano	1.14	0.109	0.351	0.0334	2.82
		-1.10	-0.119			
	Hybrid	1.717	0.108	0.360	0.0326	2.828
		-0.975	-0.118			
	Reddy	1.0356	0.1028	0.2724	0.0348	2.6411
	Present	1.13	0.106	0.350	0.0325	2.829
		-1.08	-0.121			
10	Pagano	0.726	0.0418	0.420	0.0152	0.919
		-0.725	-0.0435			
	Hybrid	0.709	0.0429	0.428	0.0151	0.921
		-0.707	-0.0448			
	Reddy	0.6924	0.0398	0.2859	0.0170	0.8622
	Present	0.718	0.0410	0.417	0.0151	0.917
		-0.717	-0.0435			
20	Pagano	0.650	0.0294	0.434	0.0119	0.610
		-0.650	-0.0299			
	Hybrid	0.653	0.0298	0.450	0.0118	0.611
		-0.646	-0.0304			
	Reddy	0.6407	0.0289	0.2880	0.0139	0.5937
	Present	0.647	0.0291	0.431	0.0119	0.607
		-0.647	-0.0298			
	CLT	±0.623	±0.0252	0.440	0.0108	0.503

Example 2. Bending of a Long Laminated Strip

The three-layer laminated strip with fibre orientation [0,90,0] is analyzed again to verify the accuracy of partial hybrid multilayer element. Similarly, the strip is supposed to be infinitely long in the y direction and simply supported along the two edges x=0 and L (see figure 46). It is subjected to sinusoidal transverse load on the top surface

$$q(x) = q_0 \sin\left(\frac{\pi x}{L}\right) \tag{5-10}$$

The lamina material properties are the same as that of example 1 in this section. They are given by equation (5-8).

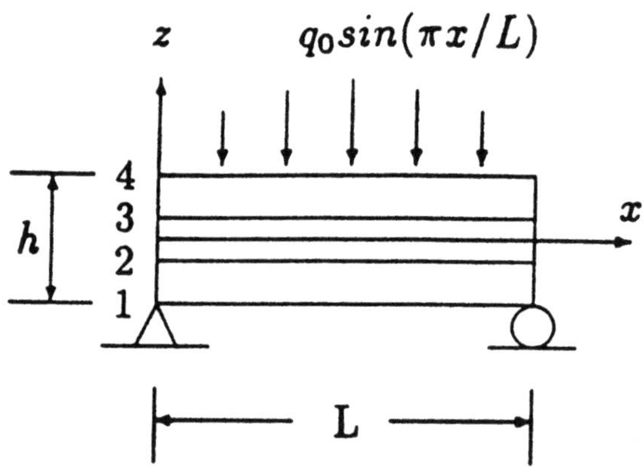

Figure 46 The cross section of infinitely long laminated strip [0,90,0] subjected to distribution transverse loading

A slice was taken out from the laminated strip for establishing the finite element model. Due to the symmetry, finite element analysis is carried out over the half of the slice. There are ten uniform multilayer elements in the half along the x-direction and one element in the y direction. Each multilayer element is composed of 12 3-D, 8-node sub-elements. The numerical results are presented in terms of normalized values which are defined as

$$\overline{\sigma_x}=\frac{\sigma_x}{q_0} \qquad \overline{\tau_{xz}}=\frac{\tau_{xz}}{q_0}$$

$$\overline{w}=\frac{100E_T h^3 w}{q_0 L^4} \qquad \overline{z}=\frac{z}{h} \tag{5-11}$$

Table 25 Maximum transverse central deflection w

S	No.	Exact	Hybrid	Present	CLT
4	4	3.023	3.022	3.029	0.510
	3	2.925	2.931	2.923	0.510
	2	2.864	2.868	2.862	0.510
	1	2.839	2.849	2.838	0.510
10	4	0.934	0.933	0.933	0.510
	3	0.933	0.9332	0.932	0.510
	2	0.931	0.931	0.930	0.510
	1	0.929	0.927	0.929	0.510
50	4	0.527	0.527	0.527	0.510
	3	0.527	0.527	0.527	0.510
	2	0.527	0.527	0.527	0.510
	1	0.527	0.527	0.527	0.510

No. : interface number.
Exact: Pagano's elasticity solution [5.16].
Hybrid: hybrid finite element method [5.25].
Present: current partial hybrid multilayer element.
CLT: classical lamination theory.

The maximum central transverse deflection $\overline{w}$ with respect to S is shown in table 25, where the surface number indicates the location of each interface of the laminate. The distribution of transverse stress $\overline{\tau_{xz}}$ and in-plane stress $\overline{\sigma_x}$ through the thickness are shown in figures 47 and 48, respectively. The results of Pagano's elasticity solution [5.16] are also given in the figures. From the figures, excellent agreement with exact solution is found for partial hybrid multilayer element method. The results of classical lamination theory are accurate only for the thin plate.

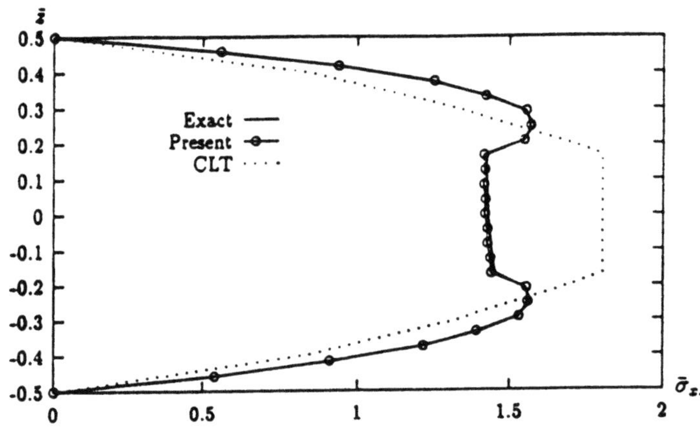

(a) S=4

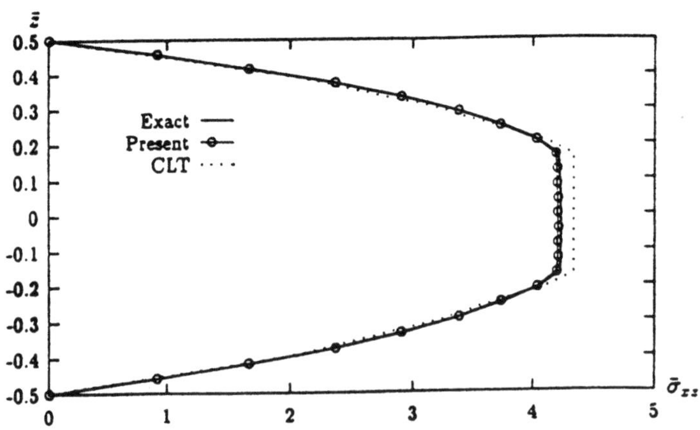

(b) S=10

Figure 47 Stress $\overline{\tau_{xz}}$ (x=0) along the thickness

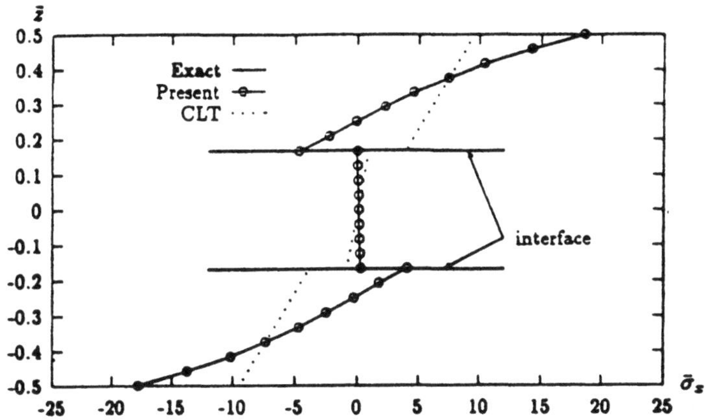

(a) S=4

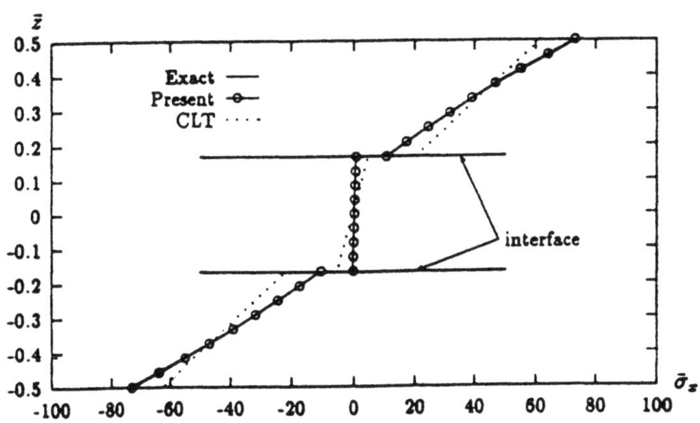

(b) S=10

Figure 48 Stress $\overline{\sigma_x}$ (x=l/2) along the thickness

The performance of finite element analysis costs 2 minutes 57.54 seconds by multilayer element solution and uses 1452 DOF, and 24 minutes 2.87 seconds by conventional displacement element solution which use 15279 DOF on VAX 6510 computer.

5.5 FINITE ELEMENT MODELS WITH DIFFERENT MESH DENSITIES

Example 1. Free Edge Effects in An Angle-ply Laminated Strip [5.19]

In the analysis of engineering structures, the state of stress within each lamina of a laminate is assumed to be planar, wherein the interlaminar stress components is neglected. However, the interlaminar stresses will appear near the free edge of a laminate and will cause delamination in the laminate. In this example, the free edge effect in an angle-ply laminated strip will be investigated.

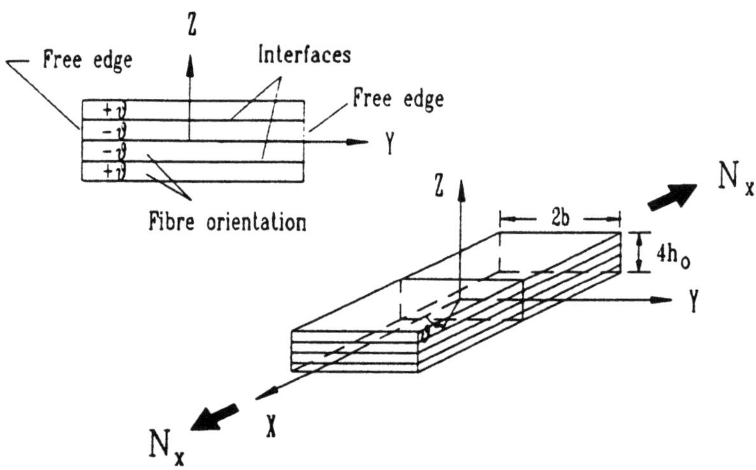

Figure 49 The laminated strip subjected to axial loading

The laminated strip to be analyzed is a finite width symmetric angle-ply laminate $[45,-45]_s$ subjected to axial tension load in plane. The loading is simulated by prescribed uniform in-plane normal strain ε_x. The elastic material properties with respect to principal material axes of each layer are

$$E_L=137.93 \text{ GPa} \qquad E_T=14.48 \text{ GPa} \qquad \upsilon_{LT}=\upsilon_{TT}=0.21$$

$$G_{LT}=G_{TT}=5.86 \text{ GPa} \qquad (5\text{-}12)$$

The geometry of a sample is shown in figure 49. The thickness of each layer is denoted by h_0, the total thickness of the laminate is $4h_0$, and the width is 2b. The ratio of the width to thickness of the laminate is $b=8h_0$. The finite element mesh is shown in figure 50. The partial hybrid multilayer elements are used to establish the finite element model. In this model, each multilayer element consists of 16 sub-elements. Due to the high stress gradient to be expected near free edge, the local region near free edge is discretized using fine mesh. Three different finite element meshes are used to investigate the stress distributions in the laminate in order to verify the accuracy of the finite element models.

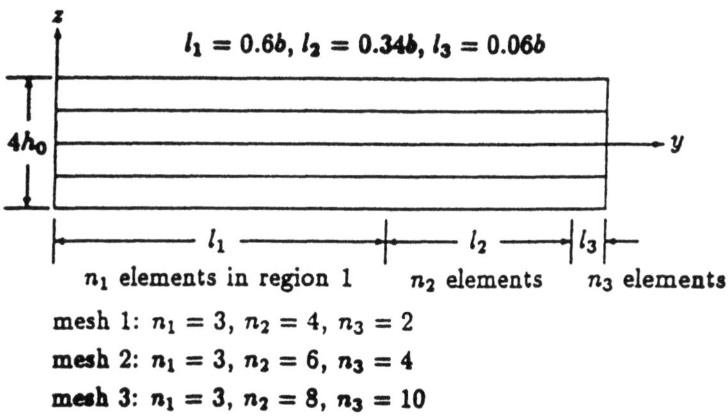

Figure 50 Finite element mesh for study of free edge effect

The results of stress distributions at the mid-plane (z=0) and at the interface ($z=h_0$) are shown in figures 51 and 52, respectively. At the mid-plane, the stresses predicted by three different meshes are the same. At the interface, the stresses obtained from three meshes are different only in the vicinity of traction free edge of the laminate.

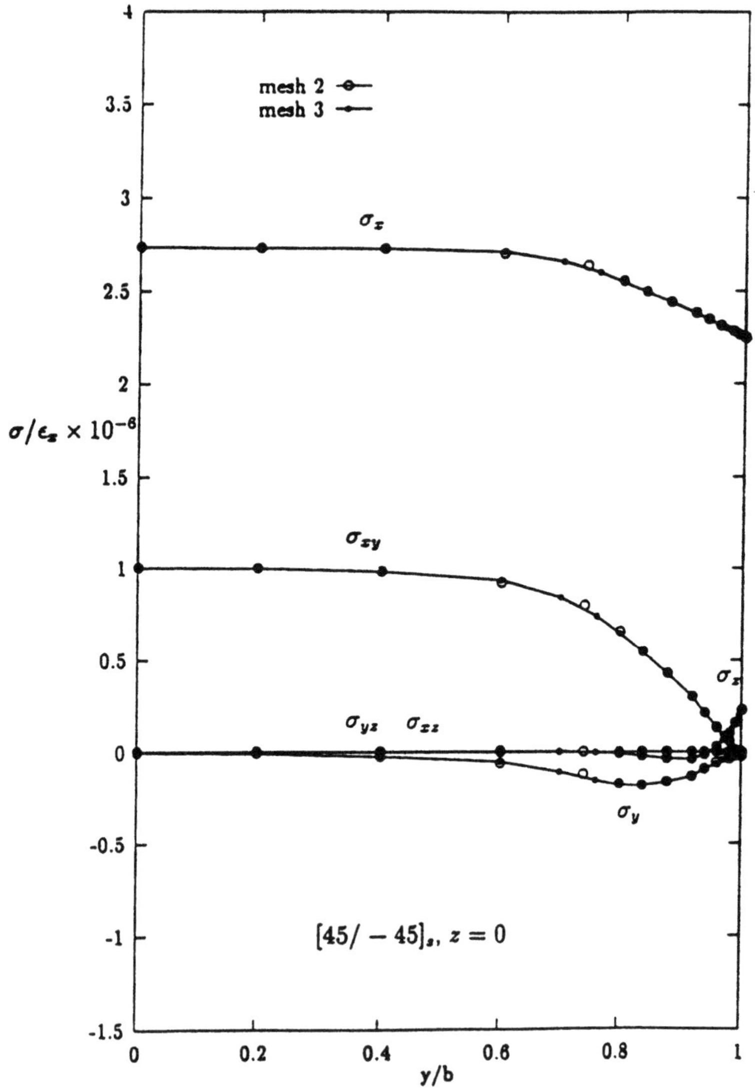

Figure 51 Stress distribution at the mid-plane z=0

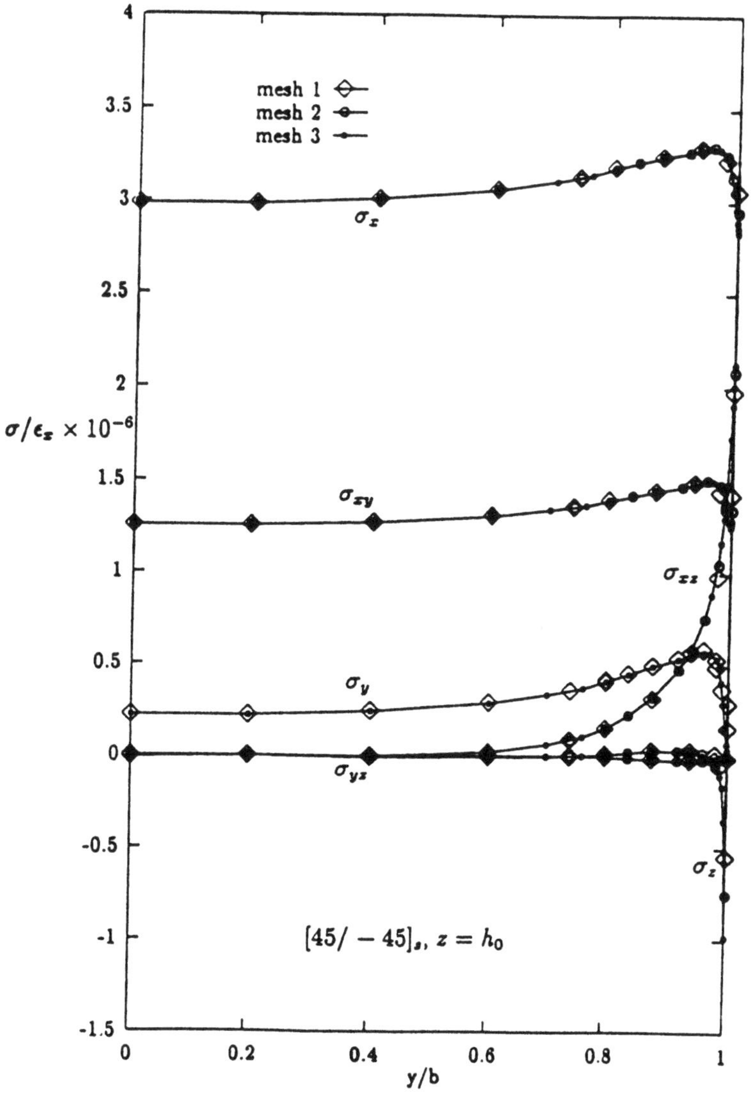

Figure 52 Stress distribution at the interface z=h_0

At the interface, the in-plane stresses σ_x, σ_y, τ_{xy} show a moderate "rise" as y/b approaches 1, but decrease to some finite value at y/b=1. The maximum value of σ_x amounts to about 10% over the average σ_x and the maximum value of τ_{xy} amounts to about 15% over the average τ_{xy} at the interface of $z=h_0$. At the free edge y/b=1, a high stress concentration of interlaminar stress τ_{xz} and a singular behaviour of interlaminar stress σ_z appear.

At the interface $z=h_0$, it is verified that the self-equilibrium conditions

$$\int_0^b \sigma_z(y,h_0)\,dy=0 \tag{5-13}$$

$$\int_0^b \tau_{yz}(y,h_0)\,dy=0$$

are satisfied.

At free edge y/b=1, the traction-free-edge conditions are

$$\sigma_y(b,z)=0 \qquad \tau_{xy}(b,z)=0 \qquad \tau_{yz}(b,z)=0 \tag{5-14}$$

From the figure 52, it is clear that the first and third conditions are satisfied exactly. Although the second condition can not be satisfied exactly at the corner ($z=h_0$, $y=b$), it can be satisfied on the average as follows

$$\int_{h_0-\varepsilon}^{h_0+\varepsilon} \tau_{xy}(y=b,z)\,dz=0 \tag{5-15}$$

in which, ε is an infinite small quantity. This is because at the corner, there are two non-zero shear stresses τ_{xz} with same magnitude but opposite sign in the two different sides of interface $z=h_0$.

Example 2. Free Edge Effects in A Cross-ply Laminated Strip [5.19]

The finite element model in example 1 can also be applied to analyze free edge effect in a cross-ply laminated strip. The laminates to be analyzed have the stacking sequences of four identical layers $[0,90]_s$ and $[90,0]_s$. The results of stress distributions σ_z, τ_{yz} and σ_y along the interface ($z=h_0$) are given in figures 53-55, respectively. These three stresses show high stress gradients in the vicinity of free edge at interface. Other three stresses are similar to those predicted with classical lamination theory ($\tau_{xy}=\tau_{xz}=0$, σ_x is almost a constant in a layer).

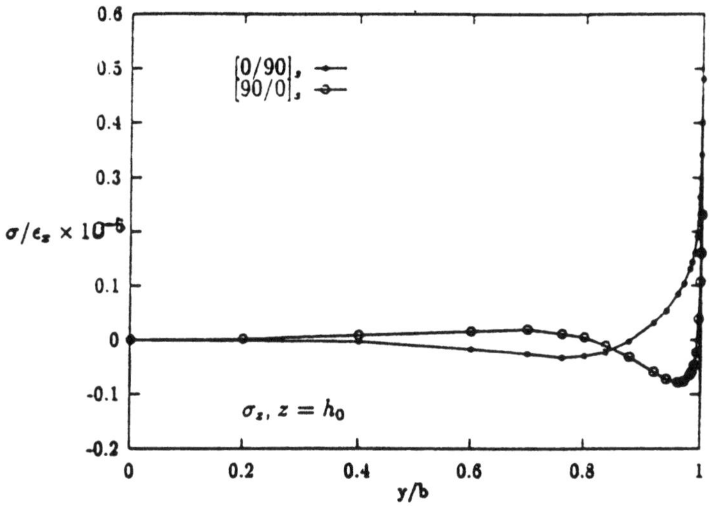

Figure 53 Interlaminar stress at the interface z=h_0

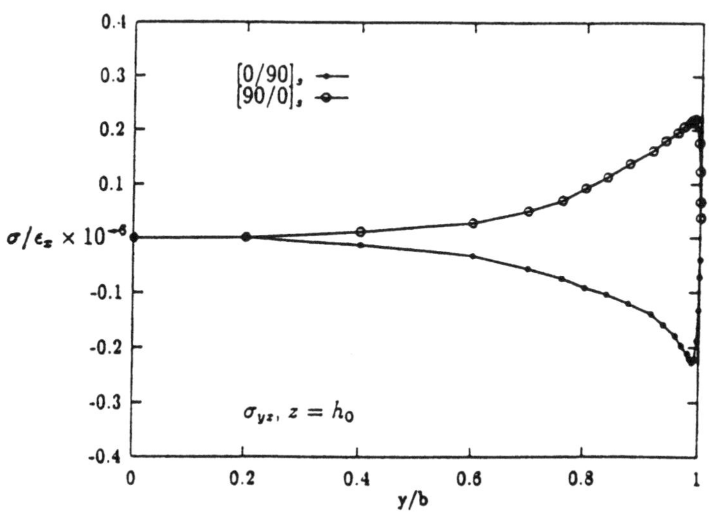

Figure 54 Interlaminar stress at the interface z=h_0

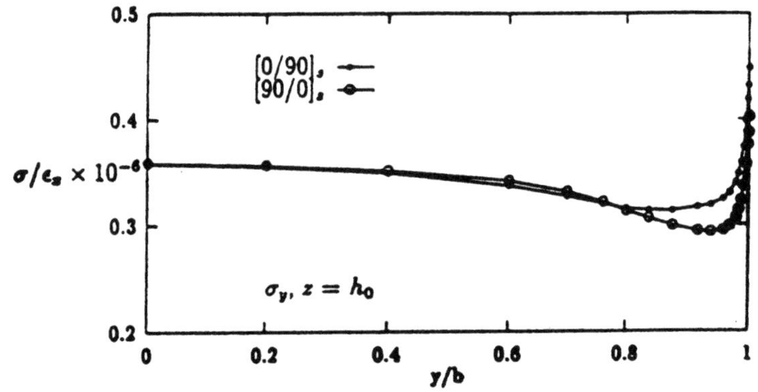

Figure 55 In-plane stress at the interface $z=h_0$

Figure 53 shows that stress σ_z rises sharply toward the free edge in a possible singular behaviour in the laminate. In figure 54, the stress τ_{yz} also increases in the vicinity of the free edge, achieves the maximum value at y/b=0.988 and then decreases to zero quickly. In figure 55, the in-plane stress σ_y rises sharply in a possible singular behaviour near free edge.

In this problem, the traction-free-edge conditions at y=b

$$\sigma_y(b,z)=0 \qquad \tau_{xy}(b,z)=0 \qquad \tau_{yz}(b,z)=0 \tag{5-14)'}$$

also must be satisfied. It can be verified that the second and third conditions are satisfied exactly in the finite element model. At the corner ($z=h_0$, y=b), there are two non-zero in-plane stresses σ_y with same magnitude but opposite sign at two different sides of the interface. Thus, the first condition can be satisfied on average meaning,

$$\int_{h_0-\epsilon}^{h_0+\epsilon} \sigma_y(y=b,z)\,dz=0 \tag{5-16}$$

5.6 GLOBAL/LOCAL APPROACH FOR STRESS ANALYSIS OF COMPOSITE LAMINATES [5.26-5.32]

As shown above, the number of degrees of freedom in a finite element model using single element type increases drastically in order to investigate stress distribution near the free edge of a laminate. In practical engineering, most composite laminates contain local regions where stress gradient is large. For example, the presence of a hole and/or free edge in laminates introduces significant transverse stresses which create a very complicated 3-D stress field in the vicinity of the hole and/or free edge. Resolution of the stress fields in these localities requires a detail 3-D finite element analysis. However, a detailed full 3-D finite element analysis of the whole laminate to obtain accurate stresses may require huge computer resources.

One way to solve these problems is to set up a global/local finite element model using different element types in different regions. It will take advantage of the properties of different elements and keep the computer storage requirement down. In chapter 4, a series of partial hybrid elements have been presented. These elements can be used in different regions in a global/local finite element model. For instance, the domain of a structure to be analyzed can be divided into three regions: local region, global region and transition region as shown in Figure 32. In the local region, 3-D solid element or multilayer element can be used to accurately determine the ply stresses in the laminate near discontinuities such as open hole, ply drop-offs, layer interface, and so on. In the global region, degenerated element can be used to predict the entire response of the structure, and the laminate is modelled as an equivalent single-layer with the smeared laminate properties. The degenerated elements are used due to simplicity and low cost. Between the global region and local region, transition elements or multilayer transition elements are used to guarantee the continuity and compatibility of displacement fields in different regions.

In this section, the effectiveness of the global/local finite element model using different element types is demonstrated by obtaining the interlaminar stresses for a laminated strip with free edge and a laminated plate with a hole. All numerical studies were performed on a VAX 6510 computer. The computational effort of each analysis is quantified by the number of degrees of freedom used in the finite element model and the computational time required to perform a stress analysis. The computational time is measured in central processing unit (CPU) time.

Example 1. Global/local Approach Using 3-D Solid Element and Transition Elements and Degenerated Elements for Analysis of Free Edge Effect

Free edge problem is an important problem in the analysis of composite laminate. The interlaminar stresses generated around the free edges and interlaminar surfaces are recognized to be the primary sources of delamination of composite laminates. Many approaches have been proposed to solve this problem. It is an ideal example to verify the efficiency and accuracy of various approaches. In this example, the global/local finite element model built by partial hybrid solid element, transition element, and degenerated element is presented to solve this problem.

The laminate with free edges to be analyzed is an angle-ply laminated strip with the [45/-45/-45/45] sequence subjected to uniaxial extension (X-direction). The laminate (shown in figure 49) has length of 2L (X-direction), width 2b (Y-direction), and thickness $4h_0$ ($W=8h_0$). Each layer in the laminate is also idealized as a homogeneous orthotropic material. The material properties are same as the example 1 in section 5.5. They are given by equation (5-12).

Because the strip is infinitely long in x direction, the displacement gradient with respect to x coordinate can be neglected and stress and strain states are independent of x coordinate. Therefore, the length of the sample to be analyzed in x direction does not affect the results of stress analysis. Thus, a slice can be taken out from the laminate to establish a finite element model. Furthermore, it can be assumed that stress distributions are symmetric about the mid-plane because the geometry, material properties, and loading are symmetric. Thus, a quarter of the slice only is needed to be analyzed.

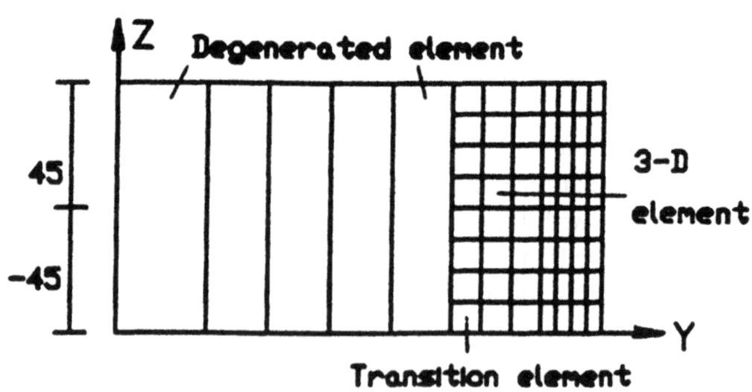

Figure 56 Global/local finite element mesh for study of free edge

The domain of the strip to be analyzed is divided into three areas: local region, global region, and transition region. The finite element mesh used for analysis is shown in figure 56. In the local region, the high stress gradient is expected. Forty eight 3-D, 20-node solid elements are used: 8 elements in thickness direction, 6 elements in y direction, and 1 element in x direction. In the central region of the laminate, five 8-node degenerated elements are used: 1 element in the thickness direction, 5 elements in y direction, and 1 element in x direction. At the transition region, eight transition elements are used to connect eight solid elements with one degenerated element. The width of the elements decreases as the free edge is approached. The problem is also analyzed by the layerwise model [5.1,5.8] and conventional 3-D displacement model. The mesh on the X-Y plane for the two finite element models is same as that for the global/local model.

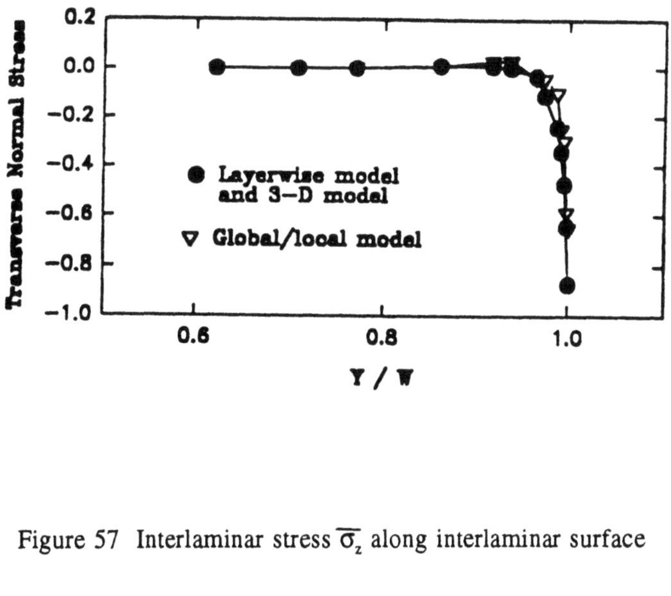

Figure 57 Interlaminar stress $\overline{\sigma_z}$ along interlaminar surface

The results of interlaminar stresses σ_z and τ_{xz} are shown in figures 57-58. The stress in the figures has been non-dimensionalized by multiplying it by the factor $20/(E_L\ \varepsilon_0)$, where ε_0 is the nominal applied axial strain of $u_0/(2L)$. The global/local model only takes 62.09 seconds CPU time on VAX 6510 Computer to solve the problem. The layerwise finite element model takes 204.40 seconds CPU time and the 3-D conventional displacement element model takes 287.06 seconds CPU time on the same computer. For the analysis, the present global/local model uses 1154 active degrees of freedom totally, the layerwise model uses 2441 active degrees of freedom, and 3-D model uses 2849 active degrees of freedom. This shows that the present global/local model takes less time and uses less active degrees of freedom than other models to solve the same problem and to get the same degree of accuracy.

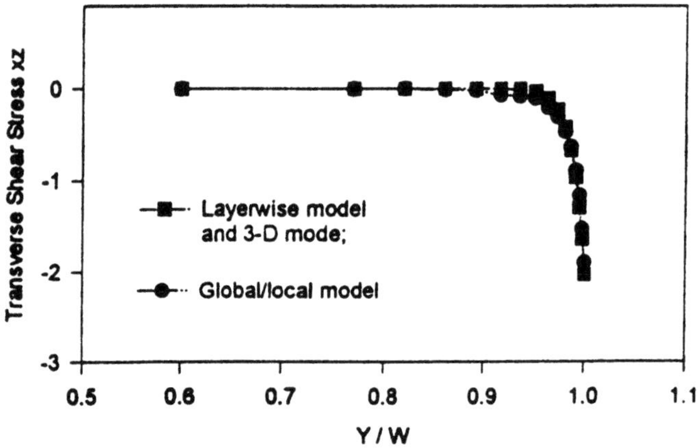

Figure 58 Interlaminar stress $\overline{\tau}_{xz}$ along interlaminar surface

Example 2. Global/local Approach Using Multilayer Elements and Multilayer Transition Elements and Degenerated Elements for Analysis of Free Edge Effect

A global/local model built by 3-D solid element, transition element, and degenerated element can predict interlaminar stresses accurately. However, it is a labour intensive task to make 3-D element mesh in the local region. In this example, 3-D solid element and 3-D transition element are replaced by multilayer elements. A global/local model built by multilayer element, multilayer transition element, and degenerated element is used to analyze the free edge problem again.

The problem to be solved is same as example 1 in this section. The domain of the strip along the Y-direction is also divided into three regions: local, global, and transition regions. The element mesh used for analysis is shown in figure 59. In the vicinity of free edge (local region), twelve multilayer elements are used along the Y-direction and each multilayer element contains 16 8-node sub-elements in the thickness of the laminate. In the central part (global region), ten 4-node degenerated elements are used in y direction. In the transition region, one multilayer transition element is used to connect the multilayer element with the degenerated element. Along the X-direction, the strip is modelled by using two elements and all elements have the same length (=L). The results of interlaminar stresses σ_z and τ_{xz} are shown in figures 60 and 61. The stresses in the figures have been non-dimensionalized by multiplying it by the factor $20/(E_L\ \varepsilon_0)$, where ε_0 is the nominal applied axial strain of $u_0/(2L)$.

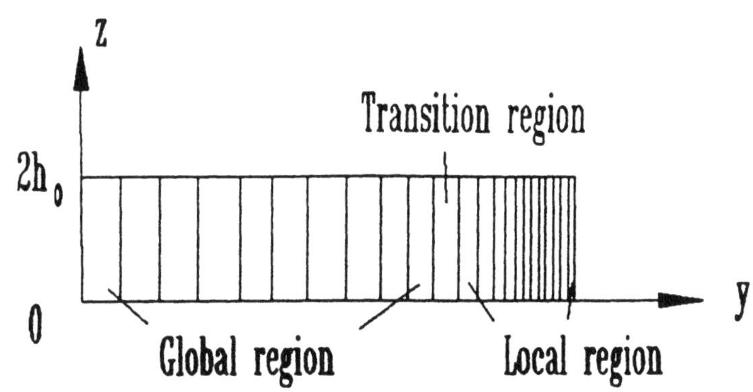

Figure 59 Element mesh for free edge problem

The problem is also analyzed by a full 3-D finite element model using 3-D 20-node solid displacement element and a previous global/local finite element model using 20-node solid elements, 8-node degenerated elements, and transition solid elements. The results of interlaminar stresses $\overline{\sigma}_z$ and $\overline{\tau}_{xz}$ calculated by the two models are also shown in figure 60 and 61. In the figures, the "previous global/local model" indicates the finite element model in example 1, and the "current global/local model" indicates the finite element model in this example. The difference of the results obtained from three finite element models is not significant. However, the computer CPU time required by three models for performance is quite different.

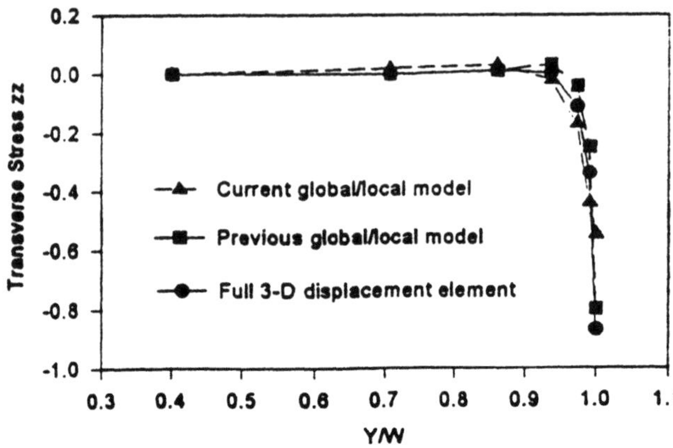

Figure 60 Stress $\overline{\sigma_z}$ at interlaminar surface along y direction

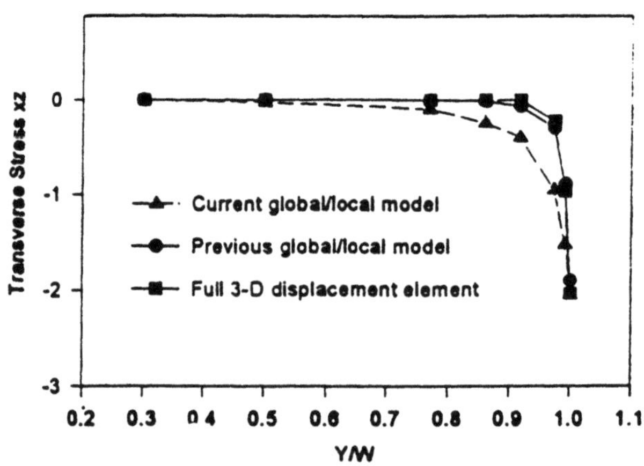

Figure 61 Stress $\overline{\tau_{xz}}$ at interlaminar surface along y direction

For the performance of the finite element analysis, the current global/local model takes 100.45 seconds CPU time, the previous global/local finite element model takes 62.09 seconds CPU time, and the full 3-D finite element model takes 287.06 seconds CPU time on VAX 6510 Computer. Therefor, the current global/local model takes about one-third computer CPU time used by the full 3-D finite element model, and the previous global/local finite element model takes about one-fifth computer CPU time used by the full 3-D finite element model. The current global/local model takes more CPU time than the previous global/local model for calculation due to the fact that during the calculation of element stiffness matrix, the current global/local model using multilayer elements must invert the matrix [H] whose size is larger than that of solid element used in the previous global/local model. In spite of that, the current global/local model is still more efficient than the full 3-D finite element model. Furthermore, the current global/local model has 2-D data structure in the finite element mesh. This is very beneficial to set up a global/local finite element model.

Example 3. Global/local Approach Using 3-D Solid Element and Transition Elements and Degenerated Elements For Analysis of A Square Laminate with An Open Hole

A square laminate $[45,-45]_s$ with an open hole is also an ideal structure to verify the efficiency of the global/local model. The stress analysis of the laminate is performed under uniaxial loading (Y-direction). The radius of the hole is R. The laminate (shown in figure 62) has length of 2L (=8R) and thickness 2h (=R). The material constants in the principal material coordinate system are given in equation (5-12).

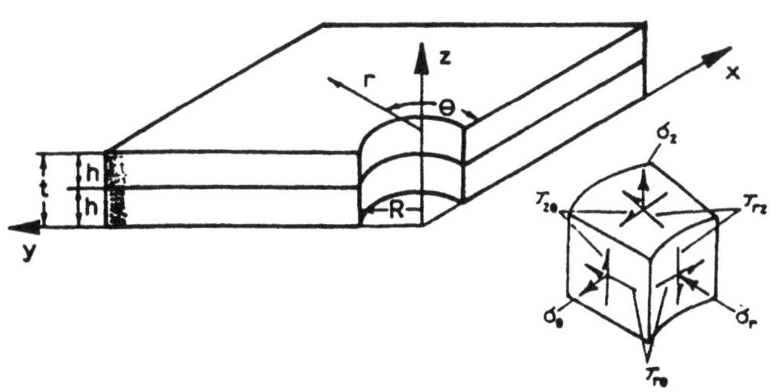

Figure 62 An angle-ply laminate with an open hole

Due to symmetry of geometry and loading in this problem, the domain to be analyzed may be reduced to one eighth portion of the laminate. In the global/local model, the domain is divided into three areas: local region, global region, and transition region (shown in Figure 63). In the local region, forty 3-D, 20-node elements are used as there is high stress gradient near the hole edge. Each layer is modeled by two 3-D, 20-node elements along the thickness. In the global region, a few degenerated elements are used. One element is used along the thickness. Between them, twenty transition elements are used to connect them. Along the thickness of the laminate, four transition elements are used to connect four 3-D solid elements with one degenerated element. The problem is also calculated by the layerwise model based on layerwise theory [5.1,5.8] and conventional 3-D displacement element model. The mesh on the X-Y plane in the two models is same as that in the global/local model.

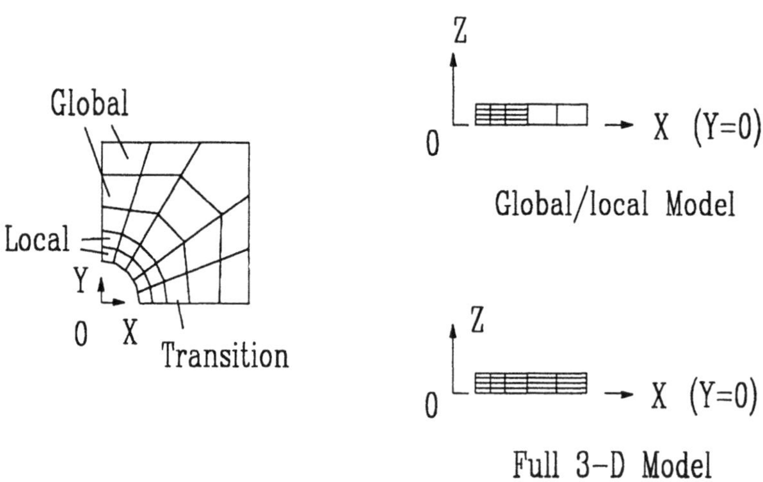

Figure 63 Finite Element Mesh for Analysis

The results of interlaminar stresses σ_z, τ_{yz} and τ_{xz} are shown in figure 64-66. The stress in the figures has been non-dimensionalized by multiplying it by the factor σ/σ_0, where σ_0 is the applied axial stress. The global/local model only takes 61.29 seconds CPU time on VAX 6510 Computer to solve the problem. The layerwise model takes 198.31 seconds CPU time and the 3-D model takes 291.17 seconds CPU time on the same computer. For the analysis, the global/local model uses 1051 active degrees of freedom totally, the layerwise model uses 2298 active degrees of freedom, and 3-D model uses 2948 active degrees of freedom. This shows that the global/local model takes less time and uses less active degrees of freedom than other models to solve the same problem and to get the same degree of accuracy.

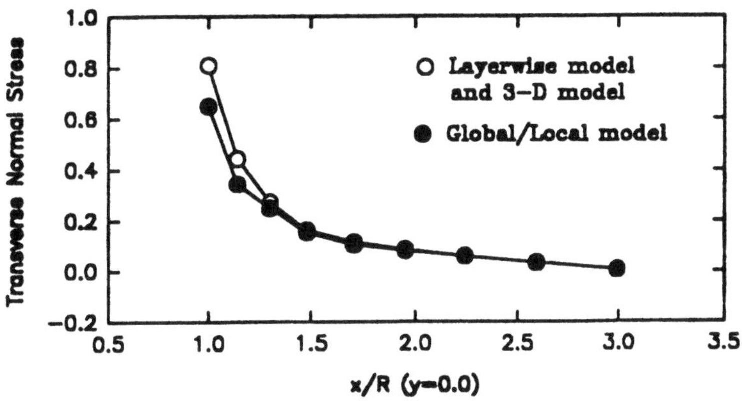

Figure 64 Interlaminar stress $\overline{\sigma_z}$ along interlaminar surface

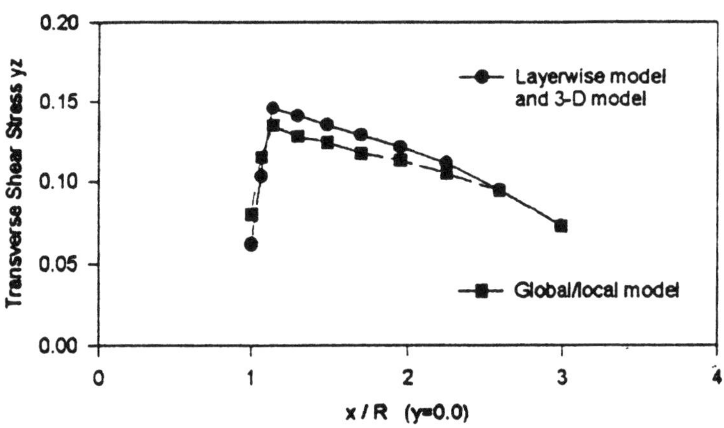

Figure 65 Interlaminar stress $\overline{\tau_{yz}}$ along interlaminar surface

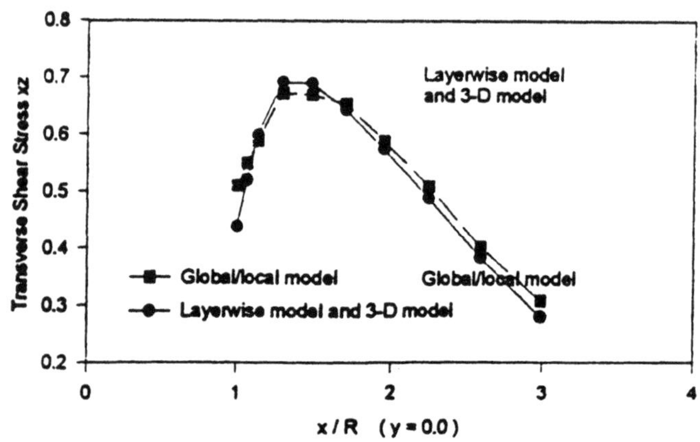

Figure 66 Interlaminar stress $\overline{\tau}_{xz}$ along interlaminar surface

5.7 CONCLUSION

In this chapter, the accuracy and efficiency of the finite element models built by partial hybrid elements have been verified. The results of the analysis show that partial hybrid elements can predict accurate stresses more efficiently than displacement elements for composite laminates. The global/local finite element approach using partial hybrid elements is more efficient than other finite element models. Once the accuracy of the finite element models built by partial hybrid elements is established, they can be applied to more general problems.

REFERENCES

5.1 S. V. Hoa and W. Feng, 'Finite elements for analysis of composite structures', in: S. V. Hoa (ed.), Computer-Aided Design of Polymer-Matrix Composite Structures, 289-359, Marcel Dekker, Inc., 1995.

5.2 I. Zeid, CAD/CAM Theory and Practice, McGraw-Hill, Inc. 1991.

5.3 O.O. Ochoa & J.N. Reddy, Finite Element Analysis of Composite Laminates, Kluwer Academic Publishers, 1992.

5.4 G.F. Carey, 'A mesh-refinement scheme for finite element computations', Comput. Meth. Appl. Mech. Eng., vol. 7, 93-105(1976).

5.5 R.J. Melosh and O.V. Marcal, 'An Energy basis for mesh refinement of structural continua', Int. J.

Numer. Meth. Eng., vol. 11, no. 7, 1083-1092(1977).

5.6 R.E. Bank, A.H. Sherman, and A. Waiser, 'Refinement algorithms and data structures for regular local refinement', in R.S. Stepleman (ed.), Scientific Computing: Applications of Mathematics and Computing to the Physical Sciences, North-Holland, Amsterdam, 3-17, 1983.

5.7 P.R. Heyliger. and Reddy, J.N. 'A higher-order beam finite element for bending and vibration problem', J. Sound Vib., vol. 126, pp.309-326,1988.

5.8 Jr. D.H. Robbins and Reddy, J.N. 'Modelling of thick composites using a layerwise laminate theory', Int. J. Numer. Methods Eng., vol. 36, pp655-677,1993.

5.9 A.K. Noor and Burton, W.S. 'Assessment of shear deformation theories for multilayered composite plates', Appl. Mech. Rev., vol. 42, no. 1, pp1-9, 1989.

5.10 C.D. Mote, Jr., "Global-local finite element", Int. J. Numer. Methods Eng., **3**, pp. 565 - 574, 1971.

5.11 A.K. Noor, "Global-local methodologies and their application to nonlinear analysis", Finite Elements in Analysis and Design, **2**, pp. 333-346, 1986.

5.12 W. Feng and S. V. Hoa, 'A degenerated plate/shell element with assumed partial stress field for the analysis of laminated composites', 4th International Conf. on Computer Aided Design in Composite Material Technology, Southampton, UK (1994).

5.13 W. Feng and S.V. Hoa, 'A partial hybrid degenerated plate/shell element for the analysis of laminated composites', Int. J. Numer. Methods Eng., vol. 39, 3625-3639(1996).

5.14 S. Srinivas and A. K. Rao, 'Bending, vibration and buckling of simply supported thick orthotropic rectangular plates and laminates', International Journal of Solids and Structures, vol. 6, pp1463-1481, 1970.

5.15 N. J. Pagano, 'Exact solutions for rectangular bidirectional composites and sandwich plates', J. Compos. Mater., 4, pp.20-34,1970.

5.16 N.J. Pagano, 'Exact solutions for composite laminates in cylindrical bending', J. Compos. Mater., 3, pp.398-411,1969.

5.17 T.H.H. Pian and M.-S. Li, 'Stress analysis of laminated composites by hybrid finite elements', in Discretization Methods in Structural Mechanics (Ed. Kuhn, G. and Mang, H.), 1989.

5.18 Q. Huang, 'Three Dimensional Composite Finite Element for Stress Analysis of Anisotropic Laminate Structures', Ph. D. Dissertation, Concordia University, Montreal, Canada (1989).

5.19 J. Han, 'Three dimensional multilayer composite finite element for stress analysis of composite laminates', Ph.D. Dissertation, Concordia University, Montreal, Canada, (1994).

5.20 W. Feng and S.V. Hoa, 'A 3-D partial hybrid laminated element for analysis of thick laminates', Third Int. Conf. on Composites Engineering, New Orleans, USA (1996).

5.21 S.V. Hoa, B. Journeaux, and L. DiLalla, 'Computer aided design for composite structures', Composite Structures, vol.13, 67-79(1989).

5.22 J.-H. Han and S.V. Hoa, 'A three-dimensional multilayer composite finite element for stress analysis of composite laminates', Int. J. Numer. Methods Eng., vol.36, 3903- 3914(1993).

5.23 W. Feng and S.V. Hoa, 'A multilayer element with partial assumed stress field for analysis of laminated structures', The 16th Canadian Congress of Applied Mechanics CANCAM 97, Quebec, Canada (1997).

5.24 J. N. Reddy, 'A simple high-order theory for laminated composite plate', J. of Appl. Mech., vol. 51, 745-752(1984).

5.25 W.J. Liou & C.T. Sun, 'Three dimensional hybrid stress isoparametric element for the analysis of laminated composite plates', Comput. and Struct., vol. 25, 241-249(1987).

5.26 S. V. Hoa and W. Feng, 'Finite element method for composites', Composites '96 and Oriented Polymers Symposium, Montreal, Canada (1996).

5.27 S. V. Hoa and W. Feng, 'Development of hybrid finite elements for stress analysis of composite structures', the 3rd International Conf. on Fracture and Strength of Solids, Hong Kong, China (1997).

5.28 S.V. Hoa and W. Feng, 'A global/local model for analysis of composites', 10th International Conf. on Composite Materials, Whistler, Canada (1995).

5.29 S. V. Hoa and W. Feng, 'Global/local approach using hybrid elements for composites', 5th International Conf. on Computer Aided Design in Composite Material Technology, Udine, Italy (1996).

5.30 S. V. Hoa and W. Feng, 'A global/local model using partial hybrid elements', XIXth International Conf. of Theoretical and Applied Mechanics, Kyoto, Japan (1996).

5.31 S.V. Hoa and W. Feng, 'Application of a global/local finite element model to composite laminates', Science and Engineering of Composite Materials, vol. 5, 151-168(1996).
5.32 W. Feng and S. V. Hoa, 'Partial hybrid finite elements for composite laminates', Finite Elements in Analysis and Design, (in press).

INDEX

MIX
Papier aus verantwortungsvollen Quellen
Paper from responsible sources
FSC® C105338

If you have any concerns about our products,
you can contact us on
ProductSafety@springernature.com

In case Publisher is established outside the EU,
the EU authorized representative is:
Springer Nature Customer Service Center GmbH
Europaplatz 3, 69115 Heidelberg, Germany

Printed by Libri Plureos GmbH
in Hamburg, Germany